MATERIALS FOR REFRACTORIES AND CERAMICS

A Study of Patents and Patent Applications

A literature study covering British, French, German and European (Munich) patents and patent applications published since 1975, compiled by J. Sigmond and edited by Marten Terpstra, technical author and researcher. The study was compiled for the Commission of the European Communities, Technical Information and Patents Division of the Directorate-General Information Market and Innovation, Luxembourg.

MATERIALS FOR REFRACTORIES AND CERAMICS

A Study of Patents and Patent Applications

Edited by

MARTEN TERPSTRA

The Hague, The Netherlands

ELSEVIER APPLIED SCIENCE PUBLISHERS
LONDON and NEW YORK

ELSEVIER APPLIED SCIENCE PUBLISHERS LTD
Crown House, Linton Road, Barking, Essex IG11 8JU, England

Sole Distributor in the USA and Canada
ELSEVIER SCIENCE PUBLISHING CO., INC.
52 Vanderbilt Avenue, New York, NY 10017, USA

WITH 16 ILLUSTRATIONS

© ECSC, EEC, EAEC, BRUSSELS AND LUXEMBOURG, 1986
Softcover reprint of the hardcover 1st edition 1986

British Library Cataloguing in Publication Data

Materials for refractories and ceramics:
a study of patents and patent applications.
1. Ceramic materials—Patents—Abstracts
I. Terpstra, Marten
016.666   TP810.5

Library of Congress Cataloging in Publication Data

Materials for refractories and ceramics.

Bibliography: p.
Includes index.
1. Ceramics—Patents. 2. Refractory materials—
Patents. I. Terpstra, Marten.
TP807.M356 1986   666   86-19797
ISBN-13:978-94-010-8422-2   e-ISBN-13: 978-94-009-4325-4
DOI: 10.1007/978-94-009-4325-4

Publication arrangements by Commission of the European Communities,
Directorate-General Information Market and Innovation, Luxembourg

EUR 10569

CONTENTS

vi

Part II: Intermediate Products and Green Bodies

INTRODUCTION

The present survey is the result of a comprehensive study of patents and patent applications submitted in France, the Federal Republic of Germany, the United Kingdom and Europe (Munich), published since 1976 and comprising more than 750 publications, classified in accordance with the INTERNATIONAL PATENT CLASSIFICATION, classes C 01 B; sub-classes 21, 33, 35; C 01 G, sub-class 25; C 04 B, sub-class 35.

The major part of the patents was granted to applicants from the United Kingdom, the Federal Republic of Germany, France, the United States of America and Japan.

Although the I.P.C. of basic refractories and ceramics produced therefrom is most useful in assessing validity and in supporting "prior art" searches, it does not satisfy the requirements of scientists and technicians who work in this field and who are also interested in the composition of such materials and the methods for producing them.

As regards Part III, relating to finished products, it was decided to follow a classification based on specific parts of machines and installations where ceramics are most commonly applied. It should be observed that ceramic machine parts and other ceramic compositions are widely spread over the I.P.C. system, comprising more than 10,000 documents published since 1975. Therefore, only applications of ceramics which belong to the C 04 35 sub-class have been discussed, but in the abstracts reference has also been made as far as possible to the ceramic composition used in producing the respective parts.

Finally, Part IV deals with apparatus for producing refractories, intermediate and finished ceramic products, again in connection with the literature classified in the C 04 B 35 sub-class.

Marten Terpstra

PART I

RAW MATERIALS

CHAPTER 1

ALUMINIUM-BASED MATERIALS

1.1 Aluminium oxides and complexes thereof

Aluminium Suisse S.A. (CH) propose in patent (12/3) a method for
producing aluminium oxide, (which contains 0.01 to 25% chrome oxide
and iron oxide) by preparing an aqueous suspension of aluminium trihydroxide, the suspension also containing soluble chrome and/or iron
components. This suspension is then exposed to a hydrothermal treatment in an autoclave at a temperature between 160 and 250°C and under
a pressure between 5 and 50 AT for a period between 20 minutes and
4 hours. During hydrothermal treatment aluminium trihydroxide is
converted into aluminium oxyhydroxide, containing 0.01-25% Cr_2O_3 and/or
Fe_2O_3.

Patent (19/1) of Asahi Glass Co.Ltd. (Japan) reveals a process for obtaining a molten refractory product, containing: 40-70 w.% ZrO_2+SnO_2;
10-58 w.% Al_2O_3+Cr_2O_3 and 2-20 w.% SiO_2. The alkali metal oxide content should be lower than that of SiO_2, while the content of SnO_2
should surpass that of SiO_2 and the content of ZnO_2 should be higher
than that of SnO_2.

According to Patent (19/2) of Asahi Glass Co.Ltd. a dense, sintered
ceramic product consists of cordierite, aluminium silicate of lithium
or aluminium titanate, furthermore earth-alkali oxides like yttrium
oxide, cerium oxide, lanthanum oxide, in an amount of 0.3 to 8 w.%.
The final ceramic product displays a low thermal dilatation coefficient
(less than 0.3%) and a porosity of less than 10%.

Patent (23/2) of <u>Babcock & Wilcox Company (US)</u> refers to the production of a thixotrope refractory furnace material of low density, containing 20-45 w.% hollow balls of aluminium oxide, with a mesh size under 4.75 mm; 0-40 w.% lamellar, sintered aluminium oxide with a mesh size under 0.3 mm and 35-80 w.% calcined aluminium oxide of a mesh size under 0.044 mm, furthermore kaolin and a polyelectrolyte type deflocculating agent.

In Patent (51/3) of <u>The Carborundum Company (US)</u> the production of an abrasive composition is described, comprising a co-fused mixture of bauxite and zirconia, which has been rapidly solidified and comminuted, the composition containing from 25 to 50 w.% zirconia, from 49.2 to 74.2 w.% alumina and from 0.8 to 2.5 w.% silica. The method of manufacture comprises: fusing a mixture of bauxite and zirconia, cooling the fused mixture to solidify the composition within three minutes of the time that cooling to solidify the composition is commenced and comminuting the resulting cooled mixture to form abrasive particles.

Patent (53) of the <u>Centre National de la Recherche Scientifique (FR)</u> concerns a mixed aluminium oxide, according to the following formula: $x^1_{(1-x)}x^2_{(x)}M^1_{(y)}M^2_{(1-y)}Al_{(11-z)}M^3_{(z)}O_{19}$ wherein: x^1 and x^2 (which may be different or identical), represent a rare-earth metal; provided that x^1 and x^2 are different, in case they represent cerium, lanthanum or gadolinium; M^1 and M^2 (identical or different) represent a metal of the group, containing magnesium, bivalent transition metals; M^3 represents a trivalent transition metal, while x, y and z (identical or different) may represent a number between 0 and 1.

According to Patent (66/7) of the <u>Compagnie Générale d'Électricité (FR)</u> elements of alkaline β -alumina (free of the allotropic β'' -variant) are produced by forming an intimate pulverulent mixture of alumina and of at least two compositions of the same alkaline metal, one of the alkaline components consisting of a carbonate or aluminate, while the other consists of fluorine, which can be in the form of a double fluoride of aluminium and an alkali metal (for example cryolite).

Patent (97/3) of <u>Ford Motor Company (US)</u> refers to stabilised suspensions of β -type aluminium oxide and of low viscosity, the suspension containing a liquid phase component, wherein a β -type aluminium oxide powder is mixed to form a finely dispersed suspension with a powder concentration of 30-70 w.%. The composition also contains a dispersion agent, composed of carboxylic acid with 3-15 C atoms and capable of chelating aluminium.

Patent (104/8) of <u>General Electric Company (US)</u> provides a process for producing a β -aluminium oxide body by preparing a suspension of β - aluminium oxide particles (with a size between 1 and 2 microns) in an organic liquid consisting of n-amyl alcohol and displays a dielectric constant of 12-24 at $25^{\circ}C$. Then a sediment is formed on an electrode by electrophoresis of the β -aluminium oxide particles, forming a compact layer on the electrode in an electric field of continuous current (100-10,000 volt/cm), followed by drying the sediment layer in air during 24 hours and by sintering at $1700-1825^{\circ}C$. The suspension may also contain aluminium stearate (0.05-0.5 w.%).

Patent (104/16) of <u>General Electric Company</u> refers to the production of an article of β -alumina-lithium, by preparing a suspension containing β -alumina particles and sodium-aluminium oxide particles (with a 14-30 w.% sodium content), whereby at least one type of the particles also contains lithium oxide. The suspension is made in an organic liquid, for example n-amyl alcohol, displaying a dielectric constant of 13.9.

Patent (108) of <u>Giulini Chemie GmbH (DE)</u> concerns a coarse-crystalline aluminium oxide, containing the following trace elements: 0.005-0.05 w.% V_2O_5; 0.005-0.5 w.% P_2O_5; 0.8-3 w.% Na_2O. Calcination of aluminium hydroxide takes place in the presence of a fluorine or vanadium salt and at a temperature between 1100 to $1300^{\circ}C$.

<u>International Business Machines Corporation (IBM),(US)</u> developed in (139/1) a process for producing β'' -Al_2O_3, the process comprising the following phases: mixing θ -Al_2O_3 with a sodium-containing component (Na_2CO_3; sodium nitrate), heating the mixture thus obtained to $950^{\circ}-1050^{\circ}C$ and maintaining this temperature for at least two hours. Na_2CO_3 is mixed with θ -Al_2O_3 at such a ratio that $Na_26Al_2O_3$ is obtained after losing CO_2.

IBM also present in Patent (139/2) a compact ceramic material with a considerable content of high-purity mullite ($3Al_2O_3-2SiO_2$) therein. The proposed initial mixture contains 3-90 w.% (preferably 5-30 w.%) mullite particles; aluminium particles in stoichiometric proportion and glass particles in an amount, necessary for the reaction, converting SiO_2 and AlO_2 into mullite form. This mixture is sintered at $1300°-1800°C$ during a period from 30 minutes to 24 hours.

A refractory composition, described in Patent (145/2) of Kaiser Aluminium & Chemical Corporation (US) consists of silicium oxide, aluminium oxide and calcined refractory clay (silex), the major part of coarse clay particles consisting of calcined refractory clay (silex), while the major part of the fine fractions is composed of silicium oxide and aluminium oxide.

S.A. Lafarge (FR) presents in Patent (166/3) a composition of CaO, Al_2O_3, SiO_2 at a 40-70% porosity, formed by pores with a ϕ between 0.6 and $8/u$ and by a structure, that consists of anorthite and aluminium oxide with a SiO_2/Al_2O_3 ratio between 2 and 0.05 and a CaO/Al_2O_3 ratio under 1.

In Patent (181/3) Montedison S.p.A. (IT) disclose a process for synthetising mordenites having a high catalytic activity, in which the SiO_2/Al_2O_3 molar ratio ranges from 16 to 40, the process comprising the step of mixing an organic base with H_2O and with at least a Na compound, an Al compound, and a Si compound, wherein the base is a diethyl-piperidinium compound, the crystallization time being longer than two days.

NGK Insulators, Ltd. (Japan) reveal in Patent (190/5) an alveolar cordierite ceramic material, which contains cordierite in the crystal phase, while containing less than 20 w.% of a spinel, mullite, corundum. More particularly the chemical composition consists of 42-52 w.% silicium oxide, 34-48 w.% aluminium oxide and 10-18 w.% magnesium. The material displays a thermal expansion coefficient of less than $22x10^{-7}(1/°C)$ in the $25°-1000°C$ temperature range.

Patent (190/6) of NGK Insulators Ltd. refers to a cordierite ceramic material, which is produced from a charge that contains raw magnesium in the form of talcum, calcined talcum, magnesite, calcined magnesite, brucite, magnesium carbonate, magnesium oxide. The average particle size of cordierite is between 21 and 100 microns. Cordierite is present in the crystal phase and displays a thermal expansion coefficient under $16 \times 10^{-7} (1/^{\circ}C)$.

Patent (191) of <u>Nikki-Universal Co.Ltd. (Japan)</u> relates to a process for producing spherical aluminium oxide particles by preparing a basic (initial) solution of aluminium chloride with an aluminium concentration of 7-12 w.% and an aluminium/chloride weight ratio of 0.3-0.8, obtained by the conversion of hydrargillite with an aqueous hydrochloride or by a hydrochloric acid solution at a temperature between 100 and $200^{\circ}C$. Next an aluminium oxide-hydrosol is obtained with a 9-15 w.% Al-concentration, this sol then being mixed with a gel-forming agent (hexamethylene tetramine, urea or their mixtures), from which gel finally the aluminium oxide product is obtained through suspension in ammonium, drying and aging.

Patent 194/3) of <u>Nippon Crucible Co.Ltd. (Japan)</u> relates to monolithic refractory material, containing a refractory mass with silica sol, aluminium oxide sol or both as a binder and a water-insoluble or nearly insoluble phosphate in form of a solid acid, which is synthesized at at $230-500^{\circ}C$. The components of the refractory mass are aluminium oxide, silicium carbide, silicium nitride, carbon and other refractory substances, which do not react with the sol binders at room temperature.

In Patent 219/1) of <u>Rhône-Poulenc Industries (FR)</u> a process is described for producing high-temperature resistant aluminium oxide agglomerates, aluminium oxide having been obtained from hydrargillite or from aluminium oxide hydrate gel, which had been dehydrated in a hot gas stream, the forming agglomerate then being processed in an autoclave in the presence of an acid. The agglomerate may also contain an oxide or lanthanum, of neodymium, preseodymium and/or thorium in an amount of 1-15 w.%. Sintering is finally carried out at $1200^{\circ}C$.

Patent (237/1) of <u>Société Européenne des Produits Réfractaires (FR)</u> refers to a fused cast refractory composition, which is suitable for use as the lining of a glass melting furnace and which comprises, by weight and on an oxide basis: 1 to 74% of CrO_3; 15 to 40% of ZrO_2;

3 to 76% of Al_2O_3; 7.5 to 20% of SiO_2; 0.4 to 2.5% of Na_2O and
0.3 to 4.0% of iron oxide and/or manganese oxide, the SiO_2/Na_2O ratio
being from 5 to 15, and the Na_2O being optionally partially or wholly
replaced by a technically equivalent amount of one or more other alkali
metal oxides, and the sum of the specified ingredients being at least
97% of the total composition.
Iron oxide is derived from a naturally occurring chromite.

In Patent (285) of K. Wolf (DE) a process is described for the production
of aluminium oxide by blowing, spraying, trickling or sucking into a
firing chamber particles of aluminium sulphate (with up to 3 mm Ø),
the temperature within the chamber being kept at 800-1250°C for a period
of time, necessary for transforming the aluminium sulphate particles
into aluminium oxide, which displays a bulk weight of 0.05-0.08 g/cm^3
and a surface of about 4 m^2/g.

Patent (287/4) of Zirconal Processes Ltd., (GB) provides a process for
producing a hydrolysable product, consisting of a fine, aqueous aluminium oxide or silica, the process comprising an etherification phase
for the implantation of alkoxy-group into the surface of the said substance. In order to promote the connection with the substance to be
implanted, the aluminium oxide may contain a hydroxyl oxide or an
oxo-bridge bond.

In Patent (237/5) Société Européenne des Produits Réfractaire reveal a
process for obtaining a composition, containing a solid solution of
aluminium oxide in silicium nitride by heating (in nitrogen atmosphere)
a material, which contains silicium oxide and aluminium oxide in mixture
with carbon. The mass of the silico-aluminous components, carbon and
a pore-forming agent (composed of particles of a ligneous substance)
is heated during a period of time, necessary for forming a product,
which contains a solid solution of β' -sialon as the only crystalline
phase.

1.2 Other aluminium-containing compounds and complexes

The British Petroleum Co.,Ltd. (GB) developed a process (42/1) for the
production of amorphous aluminosilicates and their use as catalysts

(for example for the decomposition of methanol to synthesis gas)by
reacting a source of silica, a source of alumina, a source of alkali-
metal, water and one or more polyamines other than diamine.
The polyamine may be a polyethylene polyamine having the formula:

$$H_2N \left(CH_2CHNH \right)_x H$$
$$\underset{R^1}{|}$$

where: x = an integer greater than 1 and R^1 = hydrogen or an alkyl
  group containing 1 - 6 C atoms, a cycloaliphatic group or an aromatic
group. The catalytic activity of the aluminosilicate may be enhanced
by incorporation of the metals copper, zinc, gallium, bismuth, chromium,
thorium, iron, cobalt, ruthenium, rhodium, nickel, palladium, iridium
or platinum.

Degremont S.A. (FR) present in Patent (72) a process for producing a
suspension of silico-aluminate based on a solution of sodium silicate,
by inducing a reaction between sodium silicate and a solution of aluminium polychloride (Al(OH)$_x$Cl$_{3-x}$, wherein x = 1-2,5), and agitating the
reaction mass at ambient temperature, the components' amount being so
established that a pH-value of 6-8 and a silicium content (expressed
as SiO$_2$) of 10-20 g/l can be obtained. Aluminium polychloride displays
an OH/Al ratio between 1 and 1,85.

According to Patent (82/4) of E.I. Du Pont de Nemours and Company (US)
a particulate, porous, water-insoluble amorphous poly (alumino-silicate)
is produced having a Si/Al gram-atomic ratio of about 1:1 to 10:1, a
pore volume of greater than 0.5 ml/g, an average pore diameter of 50
to 200 Å and a surface area of 200 to 600 m^2/g, by mixing an aqueous
solution of an appropriate aluminate and an aqueous solution of an appropriate silicate, allowing the aluminate and silicate in the resultant
mixture to polymerize to poly(alumino-silicate), freezing the mixture
and thereafter isolating, washing, drying and recovering particulate
poly(alumino-silicate) therefrom. The amounts, respectively, of aluminate and silicate, calculated as Al$_2$O$_3$ and SiO$_2$, are such that the total amount thereof in the mixture is about 4 to 25 w.% of the mixture,
such that the Si/Al gram-atomic ratio in the mixture is about 1:1 to
10:1, while the amount, in the mixture, of at least one water-soluble

compound, which is precipitable from the mixture at -10 to -100°C and
which is inert to the aluminate, silicate and poly(alumino-silicate) is
about 25 to 160 w.%, based on the combined weights of aluminate and
silicate, calculated as Al_2O_3 and SiO_2. The mixture is cooled to about
-10 to -100°C until it is frozen, to separate substantially all of the
chemically unbound water as substantially pure ice and to precipitate
water-soluble compound within the pores of the poly(alumino-silicate)
particles being formed.

In Patent (145/3) Kaiser Aluminium and Chemical Corporation (US) reveal
a magnesium aluminate spinel-bonded refractory, which does not exhibit
undue expansion (due to the reaction of magnesia and alumina to form
magnesium aluminate spinell) between the raw, compacted state and the
fired ceramically bonded state when fired to a temperature of 1400°C, if
a very finely divided alumina with dense particles is used. More specifically, the alumina used has an average particle size of less than
5, preferably less than 2 microns and a specific surface of less than
30 m^2/g, preferably about 5 m^2/g (i.e. the small crystallites or particles are dense and do not have a high surface area, as do finely divided active aluminas).

Patent (180/3) of Mobil Oil Corporation (US) relates to a process for
removing aluminium from aluminosilicates, having a constraint index below 12 and a major pore dimension greater than 5 nm. The process comprises: contacting the hydrogen form of the aluminosilicate at elevated
temperature (140°-760°), with an organic halide or oxyhalide containing
a halogen (chlorine, bromine, iodine) and a non-halogen component, forming an aluminium halide, depositing a non-halogen component in the
aluminosilicate, and volatilizing and removing the formed aluminium
halide.

NL Industries Inc. (US) disclose in Patent (201/6) a refractory material,
containing 45-70% aluminosilicate (mullite), 5-15% calcined aluminium
oxide, 15-35% zirconium, 1-5% bentonite and 1-5% pyrophyllite.

Patent (235/1) of SNAM Progetti S.p.A.(IT) concerns the modification of
silicium oxide by aluminium, having a porous crystalline structure, a
spec. surface greater than 150 m^2/g and corresponding to the formula:

$$1 \ Si. \ (0.0012 - 0.0050) \ Al.O_y$$

wherein: $y = 2.0018 - 2.0075$. The modification is carried out by a
process, in which a reaction is effected between an aqueous alcoholic
or hydroalcoholic solution of a silicium-derivate, an aluminium derivate
and a lattice-forming agent. The silicium derivate can be a gel of silicium oxide in combination with tetra-alkyl-orthosilicate. The aluminium
derivate is a salt, for example a nitrate or an acetate of aluminium,while
the lattice-forming agent can be chosen from among tertiary amines, amino-
alcohols, amino-acids, polyalcohols and bases of quaternary ammonium.

Patent (287/3) of Zirconal Processes Ltd. (GB) refers to a pressure moulding process wherein a dampened refractory powder of unstabilised zirconium is caused to flow, to fill the mould for the formation of a refractory brick by using a binding liquid which, on firing, leaves a refractory
residue. The binder may be based on an aluminium alkoxide or on a zirconium salt, or on an alkyl silicate.

CHAPTER 2

SILICON-BASED MATERIALS

2.1 Silicon dioxide and combinations thereof with other materials

Patent (24/1) of **BASF A.G. (DE)** describes the production of crystalline
zeolites by heating silicium dioxide and arsenic trioxide in an aqueous
solution of hexamethylene diamine to a temperature between 100 and $200^{o}C$,
followed by crystallization.

Patent (42/3) of The British Petroleum Company Ltd. (GB) refers to a
synthetic modified crystalline silica, comprising crystalline silica in
which a proportion of the silicon atoms in the crystal lattice have been
replaced by nickel. Such silica is obtained by mixing in a liquid
medium comprising either water, an alcohol, or a mixture thereof, a
source of silica, a source of nickel and an organic base in the absence
of an alkali or alkaline earth metal compound, maintaining the mixture
at elevated temperature $(300-700^{o}C)$ for a time (2-24 hrs) sufficient to
effect crystallisation of the modified crystalline silica and recovering
the modified crystalline silica so-formed.

General Refractories Company (US) disclose in Patent (105/2) a material,
based on cordierite, reaction-sintered, of high purity and low thermal
expansion coefficient; the material contains: 12-15 w.% MgO, 33-37 w.%
Al_2O_3 and 45-52 w.% SiO_2, the aggregate weight of these three components
forming 96.5% of the composition's total weight.

The disadvantage of prior art methods for producing silicon nitride
(weight losses during the reaction and uncontrolled amounts of α and
β phase materials in the end product) can be eliminated by a method
described in Patent (170/7) of Lucas Industries Ltd. (GB), the method

comprising the steps of heating a powder mixture of silica and carbon
at an elevated temperature, and contacting the powder mixture at said
temperature with a nitriding atmosphere containing silicon monoxide.
The nitriding atmosphere consists of nitrogen and silicon monoxide.

Patent (204/5) of Norton Company (US) concerns a reaction-bonded silicon
oxynitride product, having a density of 85 to 95% of the theoretical
density. The product essentially consists of 90-98 w.% of silicon oxynitride and 2-10 w.% of a glassy phase, consisting of silica and a minor
quantity of a reaction aid (an oxide of calcium, strontium, barium, magnesium, cerium, yttrium or mixtures thereof). The product displays a
modulus of rupture at $1450^{\circ}C$ of at least 5000 p.s.i. (under three point
loading using a one inch span).

A novel crystalline metal silicate with ZSM-5 structure is claimed in
Patent (228/7) by Shell Internationale Research Maatschappij (NL), the
composition of which contains oxides of hydrogen, alkali metal and/or
alkaline earth metals; silicon and trivalent cobalt, furthermore optionally
one or more oxides of a trivalent metal chosen from the group formed
by aluminium, iron, gallium, rhodium, chromium and scandium, the $SiO_2/$
$(Co_2O_3 + A_2O_3)$ molar ratio being higher than 10 and the A_2O_3/Co_2O_3
molar ratio lower than 1. The crystalline silicates can be applied as
isomerisation and dehydrogenation catalysts for paraffines.

Patent (235/2) of SNAM Progetti S.p.A. refers to a silicium oxide-based
synthetic material, consisting of crystals and modified by chemical substances, which are introduced into the crystalline structure of silicium
oxide in order to replace silicium therein. The synthetic material is
represented by the formula:

$$0.0001 - 1\ M_nO_m.1\ SiO_2$$

wherein: M_nO_m = an oxide of one or more metals used as modifying agents,
which display a more or less amphoteric character, like Cr, Be, Ti, V,
Mn, Fe, Co, Zn, Zr, Rh, Ag, Sn, Sb, S.

Patent (275/2) of Vereinigte Grossalmeroder Thonwerke (DE) reveals a refractory material, composed of a mixed, shaped, dried and fired mass of

an initial material, an additive and a binder, the initial material containing chamotte, sillimanite, corundum, mullite, zirconium silicate,
SiC or a combination thereof, the additive consisting of quartz or quartz
glass. The dried blanks obtained from this composition are fired at a
temperature between 1000° and $1500^{\circ}C$.

The proposed composition displays improved resistance to temperature
changes.

2.2 Silicon carbides and mixtures thereof

The sintered ceramic body, described in Patent (51/20) of The Carborundum
Company (US) contains about 91-99.85 w.% of silicium carbide (95 w.% of
which is in α -phase); 5 w.% of a carbonised, organic material; 0.15 -
3.0 w.% of boron, 1.0 w.% of supplementary carbon. The sintered ceramic
body of α -silicium carbide displays an equiaxial microstructure and a
volumetric mass of at least 2.40 g/cm^3.

Another Patent (51/21) of The Carborundum Company provides a ceramic composition (SiC) which is suitable for injection moulding and which contains
65 - 90 w.% of a particulate metal carbide; 12 - 30 w.% of a thermoplastic
resin, with a volatilisation temperature between 200° and $500^{\circ}C$; 2-8 w.%
of an oil or wax with a volatilisation temperature lower than that of the
resin and 0.1 - 3.0 w.% of an organo-titanate, which contains carbon,
phosphite, phosphate and pyrophosphate radicals. More particularly, the
organo-titanate can be a titanate of tetra (2-ethylhexyl).

According to Patent (51/24) of The Carborundum Company sintered ceramic
bodies are prepared, which contain 91 - 99.35 w.% of silicium carbide,
0.5 - 5.0 w.% of a carbonised organic material; 0.15 - 3.0 w.% of boron
and max. 1.0 w.% of supplementary carbon. The silicium carbide is of
amorphous character or it contains a crystalline, non-cubic α -carbide.
The specific surface of the silicium carbide varies between 7 and 15 m^2/g
and its specific weight is about 2.40 g/cm^3.

A silicium carbide composition produced in accordance with Patent (51/26)
of The Carborundum Company contains 5 - 100% of the crystalline α -phase
and (in case they are present) the following components: 2.00 w.% of

SiO_2; 0.25 w.% of free silicium; 0.50 w.% of iron; 0.50 w.% of alkali
and earth-alkali metals; 3.75 w.% of metal oxides. The particle size
of the silicium carbide powder varies between 0.10 and 2.50 microns.
The composition also contains a compacting agent, chosen from the group
of boron, beryllium (in an amount of 0.03-3.0 w.%).

Dense silicium carbide ceramic material can be obtained by a process
described in Patent (104/1) of the General Electric Company (US)
essentially by forming a homogeneous dispersion of: silicium carbide
powder (submicron size) boron or boron carbide and carbon; compressing
the dispersion in inert (argon) atmosphere at $1900^{\circ}-2000^{\circ}C$, under a
pressure of 150-700 kg/cm^2 during 10-60 minutes.

Patent (123/1) of Hiroshige Suzuki (JP) relates to finely divided silicon
carbide with a specific surface area of at least 5 m^2/g, and having a
high content of 2H-type silicon carbide, obtained by a method comprising:
reacting with a silicon source a mixture of carbon powder, containing
finely divided particles of an organic compound which is converted into
finely divided carbon at a temperature of $1200-1500^{\circ}C$ under a reduced
pressure of not lower than 10 mm Hg in the presence of metallic aluminium
or an organic or inorganic aluminium compound which is converted into
aluminium at high temperature, dispersed homogeneously in the reaction
system, the amount of the metallic aluminium or aluminium compound, calculated as aluminium, being 1-20 parts by weight based on 100 parts by
weight of the silicon carbide to be formed.

According to Patent (124/1) of Hitachi Ltd. (JP) sintered silicon carbide
with a density of more than 95% is produced by hot pressing silicon carbide powder, doped with 1 to 10 w.% of at least one element selected from
elements of the first to the fourth periods, groups IIa, IIIb, VIa and
VIII of the periodic table as an additive at 1900° to $2050^{\circ}C$ under a pressure of 100 kg/cm^2 or higher in vacuum or in an inert gas atmosphere.
The dopant, for example Al, B, Be, is incorporated in the silicon carbide
by addition to the raw materials used in the production of the silicon
carbide by electric heating.

Mannesmann A.G. (DE) disclose in Patent (174/1) a refractory ceramic
material having a base comprising carbon and a binding agent, and including 5 to 15 w.% of at least one of calcium silicon (met.), ferrosilicon (met.), and boron nitride or aluminium oxide, zirconium oxide,
and silicon dioxide.

2.2.1 β -Silicon carbides

General Electric Company developed in Patent (104/7) a process for producing a powder of silicium carbide (particle size: smaller than 1 micron),
of which a dense body can be formed from a homogeneous β -silicium carbide, containing 0.2-1.0 w.% of boron and 0.2-1.0 w.% of free carbon in
crystalline form, formed in a vapour-phase reaction. The reaction takes
place under the effect of plasma jets, by introducing into the reaction
zone a vapour, containing a silicium halogenide, a boron halogenide and
hydrocarbon. A part of the reaction zone is heated to a temperature
provoking thermal excitation and ionisation activation of the reaction
components.

According to Patent (162/3) of Kurosaki Refractories Co.,Ltd.(JP) flaky
SiC, mainly composed of β -SiC is obtained from sheet material of average
thickness 10-100 μm of an organic silicon polymer, containing carbon and
silicon atoms at the major skeletal component, by heating the sheet material at 1200-1800°C in a non-oxidative atmosphere, and dividing the sheet
material into flakes. Such SiC is especially utilized as the starting
material for ceramics having a laminar structure as well as for refractories.

Patent (265/6) of Union Carbide Corporation (US) concerns a process for
producing silicon carbide by the pyrolysis of branched polycarbonsilanes
in an inert atmosphere. The polycarbonsilanes are normally in solid state
and are soluble in non-protic organic solvents and can be produced directly from selected monomer systems.

Patent (267/14) of the United Kingdom Atomic Energy Authority (UKAEA) (GB)
refers to the production of dense sintered silicon carbide by sintering
at max. 2200°C a mixture of silicon carbide, boron and carbon in a carbon-

containing reducing atmosphere such as a methane containing atmosphere.
Surface silica which inhibits densification is thereby removed but without
removing fine carbon. According to the applicant, the increase in density
of the sintered product is obtained by inclusion of methane in the sintering
atmosphere.

In Patent (194/5) of Nippon Crucible Ltd.,Co. (JP) the production of β -
silicium carbide powder is revealed by a process comprising the heating
of a mixture of carbon powder (particle size: 20 μ or less) and a powder
of metallic silicium to a temperature between 800° and 1400°C, the molar
ratio carbon/silicium being established at 1/0.6 and 1/2; heating being
carried out in an oxidant atmosphere with an oxygen concentration between
0.3 and 35 vol.%, thereby initiating a spontaneous reaction between carbon
and silicium. The carbon powder may be selected from natural graphite,
synthetic graphite, coke and carbon black.

2.3 Silicon nitride and mixtures thereof

European Patent Application (18) of Asahi-Dow Ltd. (JP) relates to a method
for the manufacture of silicon nitride of improved quality by using a liquid
silicic acid or modified liquid silicic acid as a silicic substance and
carbon in a powdered form, a precursor of carbon in a powdered form, or a
precursor of carbon in the form of a solution as a carbonaceous substance,
and thermally treating these raw materials in a nonoxidative atmosphere
containing nitrogen. By this method, α -type silicon nitride can be easily
obtained. Particularly, finely divided α -type silicon nitride suitable
for use as the raw material for the production of high-strength sintered
articles, α -type silicon nitride whiskers useful as a reinforcing material
for ceramic and metallic articles, is produced according to the patent.

The European Patent Application of ASAHI Glass Co.,Ltd,(JP) (19/6),
provides a process for synthesising silicon nitride consisting of a
reaction of a silicon halide and ammonia at a high temperature. More
particularly, hydrogen and chlorine are burned (at least while the
reaction product is amorphous) in the reaction zone, where the halogen
containing inorganic silicon compound and ammonia are present, and the
reaction of these reactants is effected by the heat of combustion thus
obtained during a reaction time of 30-0.1 seconds. The molar ratio of
ammonia to the silicon halide has to be adjusted to 0.1-5, the molar

ratio of hydrogen to the silicon halide to 0.7-15.1 and the molar ratio
of chlorine to the silicon halide to 0.6-13.1.

Another Patent (19/7) of <u>ASAHI Glass Co.,Ltd.</u> describes a method for
synthesising amorphous silicon nitride, wherein silicon halide and ammonia are reacted in a reaction vessel at a high temperature in the absence
of oxygen to synthesise powder of amorphous silicon nitride, then the
powder is separated from a gas containing gaseous ammonia halide, which
has been produced simultaneously with the amorphous silicon nitride by
use of a collecting means.
A feature of the method is that a cold gas, containing therein neither
oxygen nor moisture, is directly mixed into the said gas to cool down the
powder and gas, so that both substances may be put in the collecting means
without causing the deposition or adhesion of the ammonium halide to the
inner wall of the reaction vessel and other component parts.

According to Patent (21/1) of <u>Association pour la Recherche et le Développement des Méthodes et Processus Industries (ARMINES) (FR)</u> an oxidation
resistant material can be obtained on the basis of silicium nitride by
adding to silicium, prior to sintering, 1-10 w.% aluminium powder and effecting a preliminary oxidation at about $1500^{\circ}C$ during at least 50 hours,
thereby obtaining acicular mullite under the effect of the reaction of
aluminium and oxygen, effected in ambient atmosphere. During preliminary
oxidation the components are exposed to vibrations.

<u>CERAVER S.A. (FR)</u> propose in Patent (56/2) a process for preparing modi-
fied silicium nitrides according to the general formula:

$$Si_{6-Z} Al_Z O_Z N_{8-Z}$$

wherein Z represents a number less than or equal to 2.8. The product is
obtained by a reaction between finely dispersed silicium nitride and
aluminium oxynitride, at a temperature above $1600^{\circ}C$ and in the presence
of at least one component capable of generating gaseous silicium monoxide
(for example a pulverulent mixture of silicium nitride and silicium
oxide), preferably in equal amounts.

Patent (56/5) of <u>CERAVER</u> also relates to a process for producing a sintered
material on silicium nitride basis by natural sintering in the presence of
a small amount of yttrium oxide. The initial silicium powder, mixed with

aluminium powder and yttrium oxide, is nitrided in a nitrogen-rich atmosphere, which also contains a small amount of carbon oxide. The nitrided powder product is crushed by means of aluminium balls and then set in a form and sintered.

Patent (74/2) of <u>Denki Kagaku Kogyo K.K. (JP)</u> reveals a process for producing ferrosilicium nitride through nitriding ferrosilicium with the aid of gaseous nitrogen at a high temperature. The product obtained by nitriding ferrosilicium is treated by immersion in water, after the treatment being removed therefrom and dried. More in particular the water treatment is applied to a product, obtained by the deformation of pulverulent ferrosilicium nitride, bonded by a bonding agent.

Patent (77/2) of <u>Didier Werke A.G.(DE)</u> concerns the production of porous, reaction-sintered, molded bodies on the basis of silicium nitride, through nitriding molded bodies, composed of silicium metal particles and additives. The initial material comprises a mixture of 70-95 w.% ferrosilicium nitride powder and 30-5 w.% silicium metal powder, with particle sizes between 200 and 40 /um.

In another Patent (77/3) <u>Didier Werke A.G.</u> disclose a process for producing sintered refractory material on the basis of silicium, aluminium, nitrogen and oxygen, by ceramic preparation, forming, sintering a mixture of silicium nitride and aluminium oxide, more particularly a mixture of 20-90 w.% ferrosilicium nitride and 10-80 w.% aluminium oxide, or a substance, which during sintering develops aluminium oxide. The mixture may also contain up to 15 w.% pulverulent metallic iron. Sintering takes place at 1600°C.

Patent (162/2) of <u>Kurosaki Refractories Co.,Ltd.</u> refers to highly durable refractory materials for steelcasting, comprising (based on the total weight of the refractory in each case) 5-30 w.% of refractory components including the principal mineral phase consisting of mullite, baddeleyite and corundum and having the chemical composition consisting of 25-85 w.% Al_2O_3, 10-70 w.% of ZrO_2 and 5-25 w.% of SiO_2, respectively, based on the total weight of the chemical composition, and 10-40 w.% of carbon powder, furthermore up to 20 w.% of one or more members selected from the group consisting of SiC, Si_2N_4, metallic silicon and ferrosilicon,

up to 30 w.% of fused silica and up to 60 w.% of alumina powder containing
at least 70 w.% of Al_2O_3 based on the total weight of said alumina powder,
and using a binder (resin, pitch), these components being moulded and the
resulting moulded product being burnt in a non-oxidating atmosphere.

Patent (170/5) of Lucas Industries Ltd. (GB) relates to ceramic materials
and to methods of producing ceramic materials.
A ceramic material, according to the patent, includes a single phase
silicon aluminium metal oxynitride, having a crystal structure based on
that of β -phase silicon nitride but having increased unit cell dimensions.
A method of producing the ceramic material, comprises; mixing silicon nitride
powder with a metal aluminium spinel powder and sintering the mixture to
form a ceramic material including a single phase silicon aluminium metal
oxynitride having a crystal structure based on that of β -phase silicon
nitride but having increased unit cell dimensions.
Preferably, the ceramic material contains at least 90% of the single
phase silicon aluminium metal oxynitride, while the said metal is lithium,
magnesium, manganese, copper, zinc, cobalt, nickel or iron.
Preferably, pressure should be applied to the mixture during the sintering
operation.

Another Patent (170/6) of Lucas Industries Ltd. describes a method of
producing a ceramic material, wherein a mixture, containing particulate
silicon nitride and a further particulate refractory material, is sintered
at an elevated temperature and in a protective environment, the further
particulate material chemically combining with the silicon nitride at the
elevated temperature to produce the required ceramic material, and wherein the mixture also contains carbon at the said elevated temperature.
The addition of carbon eliminates the problem, which exists with known methods of this kind, since silicon nitride particles inherently have a
coating of silica, which tends to inhibit the reaction between the silicon
nitride particles and the further particulate refractory material. It has
however been found that this problem can be minimised if carbon is present
at the elevated temperature of the sintering operation. It is believed
that the carbon reduces the silica on the silicon nitride particles to
gaseous silicon monoxide so that the silicon nitride particles are then
able to chemically combine with the further particulate material.
The elevated temperature is in the range $1200^{\circ}C$ to $2000^{\circ}C$ and the component, indicated as "further particulate material" consists of alumina.

A method claimed in Patent (185/4) by MTU Motoren- und Turbinen Union (DE) relates to the hot isostatic pressing of porous, silicon ceramics, wherein during pressing the compact is embedded in a powder, which (if at all) is not sinterable at process temperature.

The use of a non-sinterable powder held in a container, makes it possible to achieve the requisite gas-tight condition while preventing intimate union between the ceramic compacts and the pressure-transmitting powder. As a result the powder can be removed completely, after pressing, simply by tapping the compact so that its surface will not be damaged. Preferably, the powder consists of the same material as the compact, except that the sintering agent is omitted. This definitely prevents any chemical reaction between the envelope material and the ceramic compact. Thus, the powder may consist of pure SiC, preferably in the α -phase or of graphite.

Onoda Cement Company (JP) provides in Patent (206) a process for producing silicium nitride by heating a mixture of silicium oxide and a carbonaceous substance in an atmosphere of nitrogen and/or ammonium. The silicium nitride is of the α , β or α and β type. Silicium oxide may contain quartz, while the carbonaceous substance can be one of carbon black, petroleum coke or a carbonised resin, with a particle size under 10 microns. The mixture is heated to $1300^{\circ}-1550^{\circ}C$.

Patent (221/2) of Rosenthal A.G. (DE) refers to a process for producing silicium nitride containing structural materials of a compact glossy surface and improved resistance to oxidation by mixing together ceramic, silicate-containing additives (in powder form) and silicium raw material, shaping the powder and transforming it at $1000^{\circ}-1500^{\circ}C$ in oxygen-containing atmosphere into a self-glazing surface on a silicium nitride containing body. The initial material may contain cordierite, eukryptite, spodumene in a glass phase.

Patent (227/1) granted to the Secretary of State for Defence of GB relates to a novel method of preparing a reaction-bonded silicon nitride, containing iron oxide, which was found to enhance more the strength of the reaction-bonded silicon nitride than alumina or chromium oxide. The method includes an impregnation treatment (e.g. under vacuum) for introducing iron oxide into the silicon nitride.

According to Patent (233) of <u>SKW Trostberg A.G. (DE)</u> purified ferrosilicium
nitride (with a 80-95 w.% content of silicium nitride) can be obtained by
chlorination, whereby gaseous chlorine is caused to react with pulverulent
ferrosilicium carbide, the forming iron-III-chloride being separated in
the cooled reaction vessel. Chlorination takes place at 200-600°C (pref.
300-550°C), during 20 to 80 minutes. The residual amount of iron depends
on the duration of treatment and the reaction temperature.

Patent (253/1) of <u>Toyo Soda Manufacturing Co., Ltd. (JP)</u> concerns a high-
purity silicium nitride consisting of grainy crystals with at least 95%
α -phase content, at least 38 w.% nitrogen content and an average particle
size of max. 3/um.
Silicium nitride is obtained by heating a nitrogen-containing silane
composition to 1600°C (or higher) in an oven, wherby that part of the
oven which contact the silane composition is made of molybdenum, zirconium, lanthanum or their alloys (all having a smelting point above
1600°C).

According to Patent (253/2) granted to <u>Toyo Soda Manufacturing Co.,Ltd.</u>
silicon nitride is produced by heating a nitrogen-containing silane compound, such as $Si(NH_2)_4$ or $Si(NH_2)$, at a reaction temperature generally
higher than 400°C in the presence of ammonia; the reaction may be combined with calcination, applying, for ex.   temperatures of 1000-1600°C.
The product can contain less than 0.05% chlorine and more than 38% nitrogen.
The silane compound is produced by the vapour phase reaction of silicon
tetrachloride with ammonia in an inert atmosphere.

Patent (254/1) of <u>Tokyo Shibaura Denki K.K. (JP)</u> discloses a ceramic
product, produced by sintering a powder mixture, consisting essentially
of 100 weight parts of silicon nitride, 0.1 to 10 weight parts of at least
one of the oxide of yttrium, scandium, cerium, lanthanum and the metals
of the lanthanide series, 0.1 to 10 weight parts of aluminium oxide, 0.1
to 10 weight parts of silica and 0.1 to 4 weight parts of aluminum
nitride, wherein the ratio of aluminum nitride to silica (AlN/SiO_2) is
0.1 to 10. Sintering takes place between 1400 and 1900°C.

Patent (254/3) of <u>Tokyo Shibaura Denki K.K.</u> refers also to a method for
producing powder of α - silicon nitride which comprises the steps of:

adding 0.3 to 2 parts by weight of powder of carbon and 0.005 to 1 part
by weight of at least one silicon compound selected from the group consisting of Si_3N_4, SiC and silicon oxide nitride series compounds to one
part by weight (as converted to SiO_2) of a liquid silane derivative
which produces a precipitate and HCl by hydrolysis and further causes
SiO_2 to be grown by the baking of said precipitate, or the precipitate
produced by hydrolysis of the liquid silane derivatives; hydrolysing
the resultant mixture, if necessary; washing the mixture to separate a
solid component, if necessary; and baking the solid component for reduction and nitrogenisation at a temperature of 1300° to $1500^{\circ}C$ in an
atmosphere mainly consisting of a nitrogen gas or a gas of a nitrogen
compound.

Another Patent (254/10) of <u>Tokyo Shibaura Denki K.K.</u> reveals a process
for preparing silicon nitride powder, by baking a powdery mixture comprising: 1 part by weight of silica powder, or a silica-containing substance
(methyl silisicacid); 0.4 to 4 parts by weight of carbon powder, or a
substance generating carbon by baking; 0.005 to 1 part by weight of silicon
nitride powder synthesised by a silica reduction method, at a temperature
of from 1350 to $1550^{\circ}C$ in a non-oxidative atmosphere containing nitrogen.

Patent (255/1) of <u>Tokyo Shibaura Electric Co.,Ltd. (JP)</u> concerns a sinter
material on the basis of silicium nitride, containing crystalline bonds
formed by silicium nitride and at least an oxide of an element of group
III/a of the A.S. More particularly, the crystalline bonds consist of
silicium nitride and yttrium oxide according to the formula:
$Si_3N_4 \cdot Y_2O_3$.

Patent (276/1) of <u>Volkswagenwerk A.G.(DE)</u> describes the production of an
element, which in a first phase consists of a non-oxide ceramic material
and in a second phase of a material permitting soldering and welding, this
element being obtained by feeding in a press form pulverulent substances
for composing both phases, with the formation of a transition zone therebetween and exposing the charge to hot pressing. The ceramic material
may be composed of Si_3N_4, the press form also being added Al_2O_3, while
in the transition zone a composition of sialon type and a nitride is
provided.

2.4 Silicates

Patent (22/5) granted to Bayer A.G. (DE) refers to the production of an
alkali metal silicate (free from phase shift) according to the formula:

$$Na_{1-x}Li_xCr^{III}(Si_2O_6)$$

where x = 0–0.25 and x = 0.6–1, by preparing a mixture (in approx. stoichiometric amounts) of water-soluble or finely dispersed chrome compounds
(CrO_3, Cr_2O_3 and/or $Cr(OH)_3$), alkali and SiO_2 containing compositions as
the case may be, in the presence of reduction agents (alcohol, aldehyde,
hydrazine), under addition of water, this mixture being intensively agitated and calcined at a temperature between 1000° and $1150^\circ C$.

Patent (34) of D.H. Blount (US) reveals   a process for the production
of an aqueous dispersion of poly(organic-polysulfide-silicate) copolymer,
according to which an alkali metal hydroxide, sulphur, and an oxidated
silicon compound are mixed, then heated to just above the melting point
of sulphur while agitating for 10 to 30 minutes, thereby producing an
alkali metal polysulphide silicate; then adding it to an aqueous solution
containing an emulsifying or dispersing agent and reacting it with a
polysubstituted organic compound, thereby producing a poly(organic-
polysulphide-silicate) copolymer.

Patent (121/3) of Henkel K.G. (DE) relates to the continuous production
of finely dispersed, water-soluble potassium containing sodium-alumino-
silicates according to the formula:

$$x_1 Na_2O \cdot x_2 K_2O \cdot Al_2O_3 \cdot y SiO_2,$$

wherin: x_1 = a number between 0.2 and 1.4; x_2 = a number between
0.005 and 0.5; y = a number between 1.5 and 10; the sum of x_1 and x_2
may be between 0.3 and 1.5 (but nearer to 1); the components of a
particle size over 25 /um being present in an amount less than 0.1
w.%. The components are mixed together, and after reaction the reaction
product is filtered.

Lesieur Afrique Casablanca S.A. (Morocco) developed a process (168) for
producing a crystallised ammonium metasilicate from a filtration juice,
by decolorising such juice by passing it through a gel layer of silicium

oxide and by boiling it in order to obtain a high-concentration solution, which then is treated with a small amount of active carbon, in hot state, followed by filtration and dry evaporation of the filtrate.

By a process claimed in Patent (169) of S.F.Luca (DE) expanded silicate can be prepared, in which compact or partially expanded minerals containing water - (H_2O +) - and/or iron-III-oxide are treated with selectively oxidating (or reducing) hot gases, in one cycle, within time intervals of 0.1 to 5 seconds, at overpressure. The iron-III-oxide can be combined with pumice or perlite.

Patent (189/4) of the National Research Development Corporation (GB) discloses a method of synthesising sodium zinco/stanno/titano-silicate by mixing a concentrated aqueous sodium zincate (stannate, titanate) solution with a source of silica (e.g. powdered glass or sodium metasilicate solution) in the ratio 60g sodium silicate to sodium zincate equivalent to 8.1g zinc oxide, keeping the temperature at $40°C$ for $\frac{1}{2}$ hours, diluting threefold with water of $20°C$, allowing a precipitate to form overnight, and filtering and drying the residue at $110°C$. The resulting sodium zincosilicate may be used in its own right as a water softener or may be converted by cation exchange into any other desired zincosilicate.

According to Patent (195/1) of Nippon Gaishi K.K. (JP) a cordierite ceramic is produced comprising cordierite in the crystal phase, after being fired and having a mean pore size of 2-50 microns and a thermal expansion coefficient of not more than $16x10^{-7}$ ($1/°C$) over the range of $25-1000°C$, this cordierite ceramic having been produced from a batch including a magnesia raw material with an average particle size of 5-150 microns. The magnesia raw material is at least one of talc, calcined talc, magnesite, calcined magnesite, brucite, magnesium carbonate, magnesium hydroxide and magnesium oxide and the average particle size of the magnesia raw material is 21-100 microns.

In Patent (201/4) NL Industries Inc. (US) is proposed an organophilic clay-gel forming agent of high dispersion capacity in organic media. This agent is obtained by a reaction between a composition of methylbenzyl dialkylammonium (with 20-35% alkyl groups, with 18 C atoms) and a clay of smectite type of ion-exchanging capacity of at least 75 mval/100 g,

the amount of the ammonium composition being 100 to 120 mval/100 g clay,
(referring to 100% active clay)

Patent (219/2) of Rhône-Poulenc Industries (FR) discloses novel composition
of lead silicate with a molar ratio between SiO_2 and PbO higher than 2,
preferably between 3 and 5. The product displays an amorphous and iso-
trope character and nearly no fire losses between 500° and 900^oC. The
proposed composition is suitable for the production of vitrifiable
materials.

In Patent (228/6) Shell Internationale Research Maatschappij provide a
process for the separation, by filtration, of a fine-crystalline silicate
from an aqueous reaction mixture in which the silicate is present after
the crystallization. The mixture to be filtered contains at least 30 vol.%
(based on the said mixture) of an organic solvent that is miscible with
water, which organic solvent (with a boiling point under 100^oC) has been
chosen from the group formed by monohydric alcohols, ketones, sulphoxides
and cyclic ethers, which, per molecule, contain at most five carbon atoms.

Patent (252/1) of Tokuyama Soda K.K. (JP) concerns an alkali calcium
silicate according to the formula:

$$aNa_2O \cdot bK_2O \cdot cCaO \cdot dSiO_2 \cdot eH_2O$$

wherein: a = 0 or a number higher than 0; b = like a;   c = a number
between 7 and 9; d = a number between 30 and 34 and e = a number between
0 and 30; a+b should be above 0, but under or equal to 8.
The alkali calcium silicate displays an x-ray diffraction pattern with
diffraction peaks at distances (d) from 11.8 to 12.2 Å, 6 Å and 3 Å or
diffraction peaks at distances (d) from 13.0 to 13.4 Å, 6.7 Å and 3.1 Å.

2.5 Organic silicon compounds

Patent (82/5) of E.I.Du Pont de Nemours and Company (US) relates to a
process for preparing particulate, porous, amorphous water-insoluble
poly(alumino-silicate) having a Si/Al ratio of 2:1 to 100:1, of high
surface area, and in high yield, by mixing an aqueous solution of an
alkali metal aluminate and an aqueous solution of silicic acid, allowing
the aluminate and silicic acid in the resultant reaction mixture to

polymerise to poly(alumino-silicate), and thereafter removing sufficient water from the reaction mixture to permit recovery of the particulate poly(alumino-silicate) therefrom. The silicic acid applied is hydrated silicic acid of the formula $[Si_3O_5(OH)_2]_n$ wherein n is the degree of polymerisation of the silicic acid.

The polysilanes, claimed in Patent (79/2) by The Dow Corning Corporation (US) are substituted with $(CH_3)_3SiO$-groups and are useful for the preparation, in high yields, of fine grained silicon carbide ceramic materials. They consist of 0-60 mole% $(CH_3)_2Si$ units and 100-40 mole% CH_3Si units, all Si valences not satisfied by CH_3 groups or Si atoms being directed to groups $(CH_3)_3SiO$-, which siloxane groups amount to 23-61 w.% of the polysilane. They are prepared by reaction of the corresponding chloro- or bromo-methyl polysilanes with at least the stoichiometric amounts of $(CH_3)_3SiOSi(CH_3)_3$ and water in the presence of a strong acid.

Patent (79/3) of The Dow Corning Corporation also refers to pre-polymers which are aminated methylpolysilanes and are useful for the preparation, in high yields, of fine grained silicon carbide ceramic materials and silicon carbide-containing ceramics. They may be prepared by reaction of a polysilane containing from 0 to 60 mole percent $(CH_3)_2Si$ units and 40 to 100 mole percent CH_3Si units, wherein the remaining bonds on the silicon atoms are attached to either another silicon atom, a chlorine atom or a bromine atom, so that the polysilane contains from 10-43 w.%, based on the weight of the polysilane, of hydrolysable chlorine or 21-63 w.% based on the weight of the polysilane of hydrolysable bromine, with an aminolysis reagent having the general formula NH_2R, at a temperature of from 25° to $100^\circ C$ for a period of from 3 to 96 hours, in a suitable solvent, to recover a polysilane wherein there are bonded to the silicon atoms radicals having the formula:

$$-NHR$$

wherein R is hydrogen, an alkyl radical of 1 to 4 carbon atoms or phenyl, so that the polysilane contains from 14 to 60 w.% of -NHR when R is an alkyl radical, 14 to 66 w.% of -NHR when R is a phenyl radical, from 14 to 25 w.% of -NHR when R is hydrogen and either 0 to 25 w.% chlorine or 0-35 w.% bromine, all based on the weight of the polysilane.

Another Patent (79/5) of The Dow Corning Corporation describes a process of preparing a polysilazane polymer by contacting and reacting in an inert, essentially anhydrous, atmosphere, at a temperature in the range of $25^\circ C$ to $370^\circ C$ (A) ammonia and (B) chlorine-containing disilanes selected from the group consisting of (i) a chlorine-containing disilane having the general formula $(Cl_a R_b Si)_2$ and (ii) a mixture of chlorine-containing disilanes having the general formula $(Cl_c R_d Si)_2$, wherein a has a value of 1.5-2.0; b has a value of 1.0-1.5; the ratio of c and d is in the range of 1:1 to 2:1, the sum of $a + b$ is equal to three; the sum of $c + d$ is equal to three; and R in each case is selected from the group consisting of the vinyl group, alkyl radicals of 1-3 carbon atoms and the phenyl group.

Patent (163/2) of Kyoto Ceramic K.K. (JP) concerns the production of compact, sintered silicium carbide bodies, containing polycarbosilane by pulverising polycarbosilane and hot-pressing it in a hot-press mold, wherein the powder (with or without pressure) is heated in a non-oxidising atmosphere, involving its thermal decomposition and thereby forming SiC, which then is sintered under pressure (up to 250 kg/cm^2) at a temperature of 1900°-$2200^\circ C$.

Patent (217/2) of The Research Institute for Iron, Steel and other Metals of the Tohoku University (JP) concerns organosilicium compositions of high molecular weight and suitable for the production of shaped bodies. These compositions contain silicium carbon consisting of linear or cyclic polycarbon silane or a polycarbon silane, wherein the linear and cyclic components are chemically bonded. The linear polycarbon silane is represented by the following formula:

$$\left[\begin{array}{c} R_1 \quad R_2 \\ | \quad | \\ Si-C \\ | \quad | \\ R_4 \quad R_3 \end{array} \right]_n$$

wherein: R_1, R_2, R_3 and R_4 = hydrogen, alkyl, aryl, silyl or halogen, while the cyclic polycarbon silanes correspond to the formulae:

In Patent (263/3) <u>UBE Industries Ltd. (JP)</u> provide a method for producing organometallic copolymers, comprising (A) a polycarbosilane portion and (B) a polymetallosiloxane portion, in which the main chain contains metall-oxane units $+ M.O +$, where M is Ti or Zr, and siloxane units $+ Si.O +$. The organometallic copolymers are crosslinked with at least some of the Si atoms of the polycarbosilane portion (A) being bonded through oxygen to at least some of the M or Si atoms of the polymetallosiloxane portion (B). The organometallic copolymers can be produced by a process which comprises reacting (1) a polycarbosilane with (2) a polymetallosiloxane in an organic solvent in an inert atmosphere, thereby bonding at least some of the silicon atoms of the polycarbosilane (1) to at least some of the metal atoms M and/or silicon atoms of the polymetallosiloxane (2) through oxygen atoms. When the organometallic copolymers are molded and fired, shaped articles of an inorganic carbide can be obtained that have improved physical properties.

Another Patent (263/4) of <u>UBE Industries Ltd.</u> refers to a polymetallocarbo-silane having an average molecular weight of 700 to 100,000 and is derived from (i) a polycarbosilane having an average molecular weight of 200 to 10,000 and containing a main-chain skeleton, composed mainly of units of the formula:

$$+ \overset{R}{\underset{R}{Si}} - CH_2 +$$

wherein R represent hydrogen, lower alkyl or phenyl, and (ii) an organo-metallic compound of the formula MX_4, wherein M represents Ti or Zr, and X represents an alkoxy group having 1 to 20 carbon atoms, phenoxy or acetyl-acetoxy, at least one of the silicon atoms of the polymetallo-carbosilane being bonded to the metal atom (M) through an oxygen atom, and the ratio of the total number of the structural units $+ Si-CH_2 +$

to the total number of the structural units $+$ M–O $+$ in the polymetallo-
carbosilane being from 2:1 to 200:1. When molded and fired, the poly-
matallocarbosilane can provide articles consisting mainly of SiC and
MC having high mechanical strength and oxidation resistance.

UBE Industries Ltd. disclose in Patent (263/8) a method of preparing
polycarbosilanes, having high molecular weight and also having the major
structural unit of silicon-carbon linkage. The method comprises: heating
a polysilane at 50–600°C, in an atmosphere of an inert gas and distilling
out a low molecular weight polycarbosilane fraction with a mean molecular
weight of 300–600, followed by polymerising the fraction by heating it at
250–500°C in an atmosphere of an inert gas. The polycarbosilanes obtained
by the method of the invention are useful precursors of preparing silicon
carbide fibres and shaped articles.

Patent (265/5) of Union Carbide Corporation (US) describes a method for
preparing a crystalline microporous organosilicate, having a composition
(in terms of moles of oxides) of

$$R_2O:0–1.5\ M_2O: \, < 0.05\ Al_2O_3: \, 40–70\ SiO_2: \, xH_2O$$

wherein R represents a tetraethylammonium cation, M is an alkali metal
cation and x has a value of 0 to 15 depending upon the degree of hydration
of the composition. The organosilicate is prepared hydrothermally using
a reaction mixture comprising tetraethylammonium cations, alkali metal
cations, water and a reactive source of silica. The crystalline organo-
silicates are useful as adsorbents and in their catalytically active form,
as catalysts for organic compound formation.

CHAPTER 3

MATERIALS PREDOMINANTLY BASED ON ZIRCONIUM

Asahi Glass Company Ltd. (JP) propose in Patent (19/3) a refractory
product, which is rich in zirconium dioxide and which contains its
components in the following amounts: 85-97 w.% of ZrO_2; 0.1-3 w.%
of P_2O_5; 2-10 w.% of SiO_2; up to 3 w.% of Al_2O_3; max. 1 w.%
Na_2O. The composition may contain impurities in an amount less than
0.1-0.5 w.%.

Patent (52) of Centralniy Nauchno-Issledovatelskiy Institut Tchernoi
Metallurgii (USSR) concerns a process for the combined production of
metal alloys and zirconium-corundum, the melting of the initial components: zirconium concentrate, iron ore, aluminium being effected at
weight ratios of 51-69 : 9.9-16.5 : 19.8-34.8 at a temperature between
1500^0 and 2000^0C, while, prior to melting, a fluxing agent, like ferrosilicium, furthermore ferrotitanium, ferrosilicotitanium or metallic-
titanium can be added to the mixtures.

Dresser Industries, Inc. (US) describe in Patent (81/2) a ceramically
bonded, high-alumina content refractory material, made of a mass that
contains 3-30 w.% of a substance consisting of molten zirconium oxide
grains and molten zirconium oxide-alumina grains, furthermore of 1-15 w.%
chrome oxide, the rest of the mass comprising a material of high-alumina
content.

Feldmühle A.G. (DE) developed a process (94/2) for producing sintered
material of hard non-metal substances, e.g. carbides, nitrides, borides,
oxides of aluminium of high smelting point, incorporating therein an
oxide of zirconium and/or an oxide of hafnium. More particularly, the
said material contains 1-50 vol.% of zirconium and/or hafnium oxide,

in tetragonal, metastable state at ambient temperature. The material, which may also contain 0.05–0.25 w.% of magnesium oxide, displays a resistance to rupture (under flexion), above 500 ± 50 N/mm^2.

Patent (100/1) of Foseco Trading Co. (CH) refers to a composition containing zirconium powder and a smaller amount of an oxide of an earth-alkali metal or transition metal, the dissociation of zirconium being effected by plasma jets, the said oxide may be an oxide of iron (0.5–1.5% of the weight of plasma-dissociated zirconium) or an oxide of magnesium (1–3% of the weight of plasma-dissociated zirconium). The composition may also comprise refractory materials on the basis of oxides, carbonates, silicates and/or graphite.

NGK Insulators Ltd. disclose in Patent (190/7) a method of producing zirconia ceramics consisting mainly of ZrO_2 and Y_2O_3 in a molar ratio of Y_2O_3/ZrO_2 of 2/98–7/93 and consisting of crystal grains having a mixed phase, comprising a tetragonal phase and a cubic phase or having a phase comprising a tetragonal phase, the average size of the crystal grains being not larger than 2 μm; such ceramics have a high strength and undergo little deterioration in strength with lapse of time. The zirconia ceramics are obtained by mixing zirconium oxide having a crystal size of not larger than 1,000 Å or amorphous zirconium oxide with an yttrium compound in the above mixing ratio, moulding the mixture into an article, and firing the moulded article at 1,000 to 1,550°C.

Patent (262) granted to the Centralnij Nauchno-Issledovatel-skiy Institut Tchernoi Metallurgii (USSR) concerns a method for producing ferrosilico-zirconium corundum by melting together a zirconium concentrate, iron ore and aluminium in a weight ratio of 51-69 : 9.9-16.5 : 19.8-34.8, respectively, at a temperature of 1950-2000°C and separately casting the resulting ferrosilicozirconium and zirconium corundum phases, molten alumina at 1950-2000°C being added to the latter phase in an amount of 0.5 to 50% based on the weight of the zirconium concentrate prior to casting thereof.

CHAPTER 4

MATERIALS PREDOMINANTLY BASED ON BORON

Patent (28/1) of <u>De Beers Industrial Diamond Division (Proprietary) Ltd.
(South Africa)</u> aims at producing a hard material, essentially consisting
of $B_xC_yN_z$ in tetrahedron form, whereby x, y and z may have any value
greater than 1 or 1 and x = z, but not equal with y. The hard material
is obtained by exposing boron, nitrogen and carbon containing materials
to conditions, permitting the formation of the hard material, for example
under a pressure of min. 50 kbar and a temperature between 1300 and
$2000^{\circ}C.$

Patent (51/17) of <u>The Carborundum Company</u> relates to a process for producing a boron nitride product, the process consisting of four phases;
in the first phase boric acid (0.5-40 w.%) is mixed with fibres of boron-
oxide (60-99.5w.%); in the second phase the composition is shaped into
the required form; in the third phase the shaped body is heated in an
inert gas (ammonium, nitrogen, etc.) at a temperature above the fusion
point of boric acid but under the decomposition point of boric acid
fibres. In the fourth phase the product is heated in ammonium atmosphere
to a temperature, necessary for transforming boric acid and boron oxide
into boron carbide.

According to Patent (119/1) of <u>K.Hartl (DE)</u> mixed compounds are prepared
from alkali salts of boron acid and other oxo-acids (with a molar ratio
between alkali metal and boron varying from 1,01 to 1000), furthermore
oxo-acid anions, carbonate, phosphate perchlorate, sulphate and selenate.
The mixed compounds are characterised in that their absorption spectrum
displays a maximum between 700 and 750 nm.

Patent (131) of <u>Institut Fiziki Vysokikh Davleny Akademii Nauk SSSR (USSR)</u>
concerns compact materials on the basis of cubic boron nitride, the crystals
of which are bonded together by a binder containing (individually or in
mixture) the following intermetallic compositions: Ti_2Cu; $TiCu$;
Ti_2Cu_3; $TiCu_3$; Zr_2Cu; $ZrCu$; Zr_2Cu_3, $ZrCu_3$, in amounts between 10 and 33
vol.%. The compact material contains the following alloying elements:
V, Cr, Nb, Mo, Ta, W, Fe, Co, Ni, Mn. Altogether the compact material
is composed of 80-90 vol.% cubic boron nitride; 8-16 vol.% intermetallic
compositions and 2-4 vol.% allying agents.

<u>PPG Industries (US)</u> present in Patent (212/3) a process for producing a
refractory metal boride of submicron particles, by a reaction between
a boron source (boron trichloride), titanium tetrachloride, zirconium
tetrachloride and hafnium tetrachloride, a chlorinated hydrocarbon
(with 1-6 C atoms) in a reactor, under the effect of hot hydrogen
current, generated by a plasma generator.

According to another Patent (212/4) of <u>PPG Industries</u> titanium diboride
can be obtained, displaying a surface area of 3 to 35 m^2/g, whereby the
nominal cross-section of at least 90% of the titanium boride particles
is smaller than $1/u$. The particles consist of table-shaped or hexagonal
crystals with uniform dimensions and well developed surfaces.

The <u>Veesojuznŏ Nauchno-Issledovatelskii Institut Po Zaschite Metallov
ot Korrozii (USSR)</u> propose in Patent (278) a carbonaceous cast-moulding
composition, containing a carbonaceous filler, comprising a sulphide
coke with a sulphur content of from 1 to 8 w.% and a particle size of
from 0.00001 to 0.15 mm, thermoanthracite with a particle size of from
0.0001 to 0,30 mm, and boron carbide or silicon carbide with a particle
size of from 0.01 to 0.30 mm; a polymeric binder comprising a copolymer
of unsaturated aliphatic fluorinated hydrocarbons with an intrinsic viscosity of a 1% solution of the copolymer in acetone at $20^\circ C$ equal to
0.94 to 4.20 and a furfurol-acetone-phenol-pentadiol resin; a curing
agent comprising polyethylenepolyamine, hexamethylenediamine or sulpho-
salicylic acid; and an organic solvent; in proportions, percent by
weight: sulphide coke 15 to 20; thermoanthracite 20 to 30; boron carbide
or silicon carbide 5 to 10; copolymer(s) of the said hydrocarbons 5 to 25;

furfurol-acetone-phenol-pentadiol resin 10 to 40; curing agent 4 to 6; organic solvent 7 to 20.

4.1 Boron nitrides

A Patent (30) granted to the Bjelorusskji Politekhnicheskji Institut (USSR) refers to the production of polycrystalline boron nitrides and compact modifications thereof, the production process comprising: the preparation of a charge of hexagonal boron nitride, which is compressed into a blank under a pressure of 3-8 kbars; exposing the blank to a pressure between 100 and 150 kbars and a temperature between 290 and 1500° K during a period of time, permitting the transformation of min. 15 w.% of boron nitride into wurtzitoïd form, whereafter the compact material thus obtained is exposed to a pressure between 40 and 90 kbars and a temperature between 1600 and 3200°K during a period of time, sufficient for the complete transformation of boron nitride into solid modifications.

In Patent (38) Borax Français (FR) disclose the preparation of a finely dispersed material by an exothermal reaction between boron oxide and an electropositive metal like magnesium. The reaction takes place at a temperature between 1100° and 1200°C, by high-frequency induction heating and after the reaction the boron obtained is chemically treated with hydrochloric acid or fluorides in order to reduce the concentration of magnesium and oxygen therein, without affecting the dispersion of boron, thereby improving its quality.

General Electric Company developed in Patent (104/28) an improved process for making diamond and cubic boron nitride compacts by embedding within the mass of abrasive crystals, at least one partition strip, subdividing the abrasive crystal mass, which partition strip maintains the segregation between the separate portions of the abrasive crystal mass during step A and being sufficiently pliable not to resist the compaction of the abrasive crystal masses; subjecting a mass of abrasive crystals selected from the group consisting of diamond, cubic boron nitride and mixtures thereof, which mass is in contact with a source of a catalyst solvent for recrystallisation of the abrasive crystals, to conditions of temperature, pressures and time which result in a compact having intercrystal bonding between adjacent crystal grains; recovering the compact produced; and removing substantially all of the metallic phase from the compact of step B.

General Electric Company also provides in Patent (104/29) a process for simultaneously (1) cementing particles of cubic boron nitride (CBN) together, (2) bonding particles of ceramic together to form a substrate, or support layer, for the cemented CBN particles, and (3) bonding the cemented CBN particles to the substrate in which the cementing medium for the CBN particles and the ceramic substrate are both derived from a single multicomponent source comprising a mixture of ceramic particles and particles of an aluminium-atom yielding material.

Hitachi Ltd. (JP) claim in Patent (124/2) a process for transforming hexagonal boron nitride into the cubic form, under increased pressure (minimum 50 kbar) at a temperature, at which cubic boron nitride is thermodynamically stable. Prior to applying the transformation pressure, 3 w.% water is added to the initial substance in the form of an aqueous solution of an alkaline substance.

Institut Sverkhtverdykh Materialov Akademii Nauk Ukrainskoi SSR et Vsesojuznii Nauchno-Issledovatelskii Institut Abrazivov I Shlifovania (Lyanov, V. S, et al.) USSR present in (137/3) a process for producing cubic boron nitride by exposing a charge to a pressure of 40-70 kbars, at 1100-2000°C, the charge containing hexagonal boron nitride and a conversion initiator, for example an alkali or earth-alkali metal, furthermore an additive in form of a crystalline hydrate, a salt, containing sulphur (sulphates, thiosulphates, sulphites, etc.) and/or a halogen and/or nitrogen, the crystalline hydrate containing at least 5 mole crystallisation water. The charge contains the components in following amounts: 72.0-94.0 w.% hexagonal boron nitride; 5.0-20.0 w.% conversion initiator and 1.0-8.0 w.% crystalline hydrate.

Patent (250) of Thomson-CSF, S.A. (FR) concerns the production of solid, isotropic boron nitride, by placing in a chamber a support, on which boron nitride will be deposited at a temperature of min. 1600°C under a pressure of less than 2.10^3 Pa and by feeding into this chamber a current of gases of boron (1 mole), ammonium (20 moles) and an organic composition, which contains carbon, hydrogen and oxygen (for example alcohol) (9 moles). The gaseous boron is B_2H_6.

In Patent (279/1) the Vsesojuznji Nauchno-Issledovatelskji Institut Abrazivov I Shlifovania & Leningradskji Tekhnologicheskji Institut imeni Lensoveta (USSR) disclose an ultrahard material, containing cubic boron nitride and a binding agent, which consists of a carbide and diboride of a metal of groups IV and V of the P.S. with following eutectic proportions of the components: 28-80 w.% cubic boron nitride and 72-20 w.% carbide or diboride of a metal of group IV or V of the P.S.

The Vsesojuzmji Nauchno-Issledovatelskji I Proektnji Institut Tugoplavkikh Metallov I Tverdykh Splavov, (USSR) reveal in Patent (280) a polycrystalline material on the basis of diamond and cubic boron nitride in combination with complex compositions of earth-alkali metal oxides (calcium and/or magnesium oxide), the complex compositions of earth-alkali metals being also in combination with metals like boron, iron aluminium, chrome, nickel, copper and their oxides.

CHAPTER 5

LITHIUM-BASED MATERIALS

Patent (176/6) of Max-Planck Gesellschaft zur Förderung der Wissenschaften e.V. (DE) discloses a crystalline lithium nitride, obtained by heating lithium metal (of 99.9% purity) in a vessel, made of tungsten, niobium, ruthenium or tantalum, in a nitrogen containing atmosphere at a pressure of min. 250 mm Hg with total exclusion of oxygen and water traces, the heat treatment taking place at a temperature between $300^{\circ}C$ and the fusion point of lithium nitride.

Patent (202/1) of Nobutoshi Daimon (JP) concerns lithium taeniolite, having the formula $LiMg_2Li(Si_4O_{10})F$ or $LiMg_2Li(Ge_4O_{10})F_2$ (wherein X represents Si or Ge) and is obtained by mixing lithium oxide and magnesium fluoride and optionally magnesium oxide or magnesium oxide and lithium fluoride and optionally lithium oxide with silicon oxide or germanium oxide so, that the molar ratio of Li:Mg:Si (or Ge):F is 1:1:2:1.1 to 1.3, the fluoride being included in an excess amount of 10-30 molar % to compensate for the loss of fluorine during the process; melting the mixture at a temperature in the range $1,250^{\circ}C$ to $1,450^{\circ}C$; and slowly cooling until the melt crystallises. From such lithium taeniolite an aqueous sol of flake-like ultra-fine particles can be made, wherein most of the flake-like particles dispersed in water have a thickness of less than 150 Å.

CHAPTER 6

CARBIDES OF METALS OR METALLOIDS OTHER THAN THOSE
REFERRED TO IN THE PREVIOUS CHAPTERS

In Patent (27/3) The Battelle Memorial Institute (CH) claim a process
for the preparation of an ultrahard material, based on metal carbides
(tungsten and molybdenum carbides) and having a hexagonal crystal
structure identical with that of tungsten carbide. According to this
process, one heats between $1000^{\circ}C$ and a temperature T_x, which is lower
than the maximum stability limit of the $Mo_xW_{1-x}C$ phase wherein
$0.01 < x < 1$, a mixture, intimate to the molecular or atomic scale, of
tungsten and molybdenum the total content of which in Fe, Ni and Co
does not exceed 0.1% with carbon and/or a carbon compound, T_x being
defined as follows: for $0.01 < x < 0.8$, $T_x = 2700-1375x^{\circ}C$;
for $0.8 < x < 1$, $T_x = 3400-2250x^{\circ}C$.

The Carborundum Co. reveal in Patent (51/10) a sinterable pulverulent
mixture of silicium carbide powder, furthermore of 0.5 - 5.0 w.% carbon
and beryllium or a beryllium compound, in such amounts that the mixture
contains 0.03-3.0 w.% of beryllium. The composition may contain
0.03-3.0 w.% of a blend of beryllium and boron or compounds thereof.
The powder particles are smaller than 5 µm.

Another Patent (51/14) of The Carborundum Company refers to particulate
silicium carbide compositions containing at least 95 w.% silicium carbide and a sufficient excess amount of negative doping elements with
regard to the positive doping elements, to ensure that the volumetric
resistance against cold should be reduced by about 1.25 Ω.cm and that
the ratio between the volumetric resistance to cold $(20^{\circ}C)$ and that to
heat $(1200^{\circ}C)$ should be less than 12/1. The aluminium carbide consists of particles smaller than 20 µ and is compressed under a pressure

of $7.10^6 - 7.5.10^7$ Pa at a temperature of 1800-2400°C, during a period of time necessary for establishing a specific weight of about 2.5 g/cm^3 of the material.

Yet another Patent (51/15) of <u>The Carborundum Company</u> concerns the sintering of silicium carbide powders, which comprise boron or boron-containing substances, used as densifying agents in order to produce ceramic products of increased density. Sintering takes place in a boron-containing atmosphere with a partial pressure of boron amounting to at least 10^{-2} Pa. The boron content of the silicium carbide powder is about 0.1-5.0 w.% and the boron-containing atmosphere may also comprise an inert gas (nitrogen, helium).

Patent (223) of <u>E.Rudy (US)</u> provides a metal carbide, which can be used in processing hard metals and which consists of $(Mo_x W_{x'} Cr_{x''})C_z$, where the relative mole parts x, x', x" (x+x'+x" = 1) of the metal components correspond to min. 0.20 for x, to 0-0.8 for x' and to 0-0.18 for x", while the stoichiometric parameter z amounts to 0.6-0.73. The metal carbide consists of a double-phase mixture of metal subcarbide M_2C and hexagonal monocarbide MC, in form of mixed crystals or a solid solution. M represents the metal component. The average distance between the grains of the subcarbide and monocarbide phase is less than 4 µm. The double-phase mixture is obtained by the decomposition of the solid-state body of the pseudocubic η M_3C_2 or the cubic α-MC_{1-x} high temperature mixed crystals.

According to Patent (261) of <u>W.P.Tshviruk, et al. (USSR)</u> packing materials (bodies) are prepared on graphite basis and contain titanium carbide, the components being used in the following amounts: 59-35 w.% graphite and 31-65 w.% titanium carbide. The composition also comprises a carbonaceous binding agent and can be used in decomposing alkali amalgams.

In a process described in Patent (263/6) of <u>Ube Industries Ltd. (JP)</u> carbonitrides of metals of Groups IV, V and VI of the Periodic Table are prepared by calcining a precursor obtained by (i) reacting the reaction product of ammonia and the halide of a metal selected from the group consisting of Groups IV, V and VI of the Periodic Table of Elements, with

polyphenol, or (ii) reacting the reaction product of polyphenol and the
halide of a metal selected from the group consisting of Groups IV, V
and VI of the Periodic Table of Elements with ammonia, the amount of
the polyphenol being within the range defined by the following relationship:

$$0 < \frac{a \times b}{c} < 3$$

wherein a is the number of the hydroxyl group contained in one molecule
of the polyphenol, b is the number of moles of the polyphenol and c is
the number of moles of the metallic halide. The desired metallic carbonitride in the form of finely divided powder, having a uniform size and
an excellent sintering property, can be obtained at low energy consumption.

Union Carbide Corporation (US) reveal in Patent (265/1) a process of
preparing sintered metal carbide articles, essentially by impregnating
a carbohydrate substance with a metal, for example selected from tungsten,
titanium, tantalum, molybdenum, zirconium, hafnium, thorium or their
mixtures, while the carbohydrate component may consist of sugar or amidon.
The impregnated substance may also be cellulose or rayon.

Patent (274) of Vereinigte Aluminium Werke A.G. (DE) describes a process
and equipment for the continuous production of metals and metal carbides,
through thermal reduction. The process includes the agglomeration of a
mixture of a metal oxide (an oxide of boron, silicium, titanium, zirconium, tantalum, niobium, molybdenum, tungsten or uranium) and carbon,
each agglomerated lump being coated with carbon and/or graphite, whereafter the bodies thus formed are embedded in a packing to be coked and
reduced to a metal carbide. Reduction takes place by electric resistance
heating. The metal is extracted with a halogenide of the same metal.
For example, the extraction of aluminium carbide is effected by aluminium
fluoride.

CHAPTER 7

VARIOUS COMPOSITIONS MAINLY USED AS CATALYSTS OR CATALYST COMPONENTS

7.1 Catalysts

According to Patent (13) of the Aluminiumipari Tervező és Kutató Intézet and Almásfüzítői Timföldgyár (Hungary) the Bayer-process for the treatment of bauxite (containing diaspores and/or goethite and/or haematite in finely dispersed state) is accelerated, goethite is transformed into haematite, which then is recrystallised. The treatment takes place at a temperature between 180° and $300^{\circ}C$ with the application of an aluminate solution that contains 80 to 300 g/l Na_2O and a catalyst with the structural formula:

$$A_2B_3/SiO_4/_{3-x}/(OH)_4/_x$$

wherein A = Ca^{2+} and/or Mg^{2+} and/or Mn^{2+} and/or Fe^{2+} and B = Al^{3+} and/or Fe^{3+} and/or Cr^{3+}. The catalyst is added in a quantity amounting to 5-20% of the dry bauxite. The process increases the alumium yield.

Patent (42/2) of The British Petroleum Co.Ltd. (GB) refers to a process for the production of synthesis gas from methanol, which process comprises: contacting methanol in the vapour phase at elevated temperature with a catalyst, comprising a crystalline silica modified by inclusion of cobalt in the crystal lattice in place of a proportion of the silicon atoms. The atomic ratio of cobalt to silicon in the modified crystalline silica is in the range from 1:5 to 1:500. More particularly the modified crystalline silica is prepared by mixing in a liquid medium comprising water, an alcohol or a mixture thereof, a source of silicon, a source of cobalt, a nitrogenous base and, optionally, a mineralising agent and/or inorganic base, maintaining the mixture under such conditions of temperature and pressure and for a time sufficient to effect crystallisation of the modified crystalline silica, separating the modified crystalline silica product from the liquid medium and thereafter calcining the product. The nitrogenous base is an alkanolamine, a tetraalkylammonium compound in which the alkyl group contains from 1 to 5 carbon atoms or a tetraarylammonium compound in which the aryl group is a phenyl or an alkylphenyl group.

Patent (60) of <u>Chemische Werke Huls A.G. (DE)</u> concerns a process for producing crystalline metal silicates by means of linear branched polyamines with more than 2 N-atoms of the general formula:

$$\begin{array}{c} R \\ \diagdown \\ R' \end{array} N - \left[\begin{pmatrix} R \\ C \\ R \end{pmatrix}_x - \begin{matrix} R \\ N \\ R \end{matrix} - \begin{pmatrix} R \\ C \\ R \end{pmatrix}_y \right]_z -N \begin{array}{c} \diagup R \\ \diagdown R \end{array}$$

where the rests R can be combined in any way with hydrogen, C_1-C_{10}-alkyl and cycloalkyl, while x, y and z represent numbers between 1 and 10. The bonding agents are intimately mixed with the metal (pref. Al)silicate and agitated during 5-100 (pref. 5-14) days at 80-220^oC, whereafter the solidified matter is calcined at 500-600^oC, exposed to ion exchange and again calcined.

Patent (61/1) of <u>Chevron Research Company (US)</u> refers to the production of a synthetic crystalline chromia silicate catalyst having an X-ray diffraction pattern similar to known aluminosilicates (e.g. ZSM-5), by hydrothermally crystallising an aqueous reaction mixture containing sources of quaternary alkylammonium oxide, chromium oxide, silica and an alkali metal oxide. The chromia silicate has a SiO_2:Cr_2O_3 mol. ratio greater than 20:1. The crystalline chromia silicate is useful as a catalyst in hydrocarbon conversion processes, such as dewaxing and olefin production.

In Patent (76/1) <u>Deutsche Gold- und Silber-Scheideanstalt vormals Roessler (DE)</u> disclose the production of composite bodies, consisting of two or more ceramic or metallic components of different specific weight and with a continuous transition between these components. The pulverulent solid substances are dispersed in an inert medium and then sedimented under the effect of centrifugal forces.

Patent (201/2) of <u>NL Industries Inc.</u> provides a catalyst for the alkylation of aromatic hydrocarbons with a mono-olefin which comprises a trioctahedral 2:1 layer lattice smectite-type mineral containing a metallic cation having a Pauling-electronegativity greater than 1.0 in cation exchange positions on the surface of the said mineral, the said mineral being a hectorite-type clay having the structural formula:

$$[(Mg^{2+}{}_{6-x}Li^+{}_x)^{VI} Si_8O_{20}(OH)_{4-y}F_y]\overset{x}{\underset{z}{\longrightarrow}}M^\bullet$$

wherein $0.33 \leqslant x \leqslant 1$, $0 \leqslant y \leqslant 4$; a stevensite-type clay having the structural formula:

$$[(Mg^{2+}{}_{6-x})^{VI} Si_8O_{20}(OH)_{4-y}F_y]\overset{2x}{\underset{z}{\longrightarrow}}M^\bullet$$

where $0.16 \leqslant x \leqslant 0.5$, $0 \leqslant y \leqslant 4$; or a saponite-type clay having the structural formula:

$$[(Mg^{2+}{}_6)^{VI} (Si_{8-x}Al^{3+}{}_x)^{IV} O_{20}(OH)_{4-y}F_y]\overset{x}{\underset{z}{\longrightarrow}}M^\bullet$$

where $0.33 \leqslant x \leqslant 1$, $0 \leqslant y \leqslant 4$ and wherein M is at least one charge balancing cation having a Pauling-electronegativity greater than 1.0 of valence z, the smectite-type mineral being synthesised by a hydrothermal treatment of a gel containing the required molar ratios of silica, alumina, magnesia and fluoride and having a pH of at least 8 at a temperature of from 100°C to 325°C for a period of time sufficient to crystallise the desired smectite.

Patent (204/8) of <u>Norton Co. (US)</u> relates to alumina bodies with a specific surface of 70-220 m^2/g, a pore volume of 60-85%, a pressure resistance of at least 3.2 kg for pellets with 2.5 mm ∅ , an abrasion loss of less than 5 w.% and a loss, due to wear, of 0.22 w.%. The initial material contains at least 99.5% Al_2O_3. The bodies are essentially produced by processing a mixture of gibbsite and boehmite with a monobase acid, water, and firing the fluid mixture at 1200°C. The produced bodies can be used as catalyst supports.

In Patent (204/11) <u>Norton Company</u> reveals a shaped element that can be used as a catalyst, the element being composed of Al_2O_3 (of at least 99.5% purity) present in the composition in the gamma, kappa, delta, theta and alpha phase or the mixtures thereof. More particularly, the high-purity aluminium oxide compostion is formed <u>a</u>) of calcined gibbsite (30-80 parts) with a firing loss less than 8%; <u>b</u>) of microcrystalline boehmite and c) of microcrystalline, desensitised boehmite. Into this mixture also a mono-acid is introduced (2-10 w.%), furthermore water, in an amount to promote the fluxdity of the mixture, which then is shaped, dehydrated and sintered at 625° - 1200°C.

According to a process (228/1) developed by Shell Internationale Research Maatschappij a crystalline iron silicate of characteristic X-ray diffraction pattern is prepared from a base mixture in which the silicon (Si), alkali metal (M), iron and tetra-alkylammonium (R_4N) compounds are present in such quantities that the $[M_2O+(R_4N)_2O]/SiO_2$ and $M_2O/(R_4N)_2O$ molar ratios amount to 0.24-0.40 and 0.4-1.0, respectively and $SiO_2:Fe_2O_3$ ratio is 10. In combination with a methanol synthesis catalyst this material shows excellent performance as a catalyst in the preparation of hydrocarbons from synthesis gas. It may also be used alone as a catalyst.

Patent (228/2) of Shell Internationale Research Maatschappij relates to a process for the preparation and separation of paraxylene from mixtures of Ca aromatic hydrocarbons, the process comprising:
a) the separation by adsorption of the feed mixture into a mixture (I) of para-xylene and ethylbenzene and a mixture (II) of ortho-xylene and meta-xylene; b) the separation of para-xylene by crystallisation of (I), with subsequent distillation in order to obtain ethylbenzene with recirculation of the residue to the crystallisation step; c) the isomerisation of (II) after its distillation; d) the separation of para-xylene from the isomerisate by isomerisation; e) the recirculation of the components not adsorbed in step d), to the isomerisation step.
A crystalline silicate with the composition (in moles of oxides)

$$\frac{(1.0 \pm 0.3)(R)_2O}{n}(aFe_2O_3 . bAl_2O_3 . cGa_2O_3$$

$$.y(d.SiO_2 . eGeO_2),$$

wherein R = one or more mono- or bivalent cations, $a \geqslant 0.1$; $b \geqslant 0$; $c \geqslant 0$; $a + b + c = 1$; $y \geqslant 10$ - preferably less than 300; $d \geqslant 0.1$; $e \geqslant 0$ - preferably = 0; $d+e = 1$; n = valency of R is used as well as adsorbent or as isomerisation catalyst.

Patent (55) of Centro Ricerche Fiat S.p.A. (IT) discloses a monolithic support for a catalyst, suitable for use in controlling carbon monoxide emission, prepared by forming a homogeneous, fluid semi-solid mass, by admixing colloidal γ-alumina, α alumina monohydrate and ceramic fibres (e.g. aluminosilicate fibres) with fluidising and binding agents, water and a mineral acid, this acid being used in an amount sufficient to convert the α Al_2O_3 monohydrate into a gel, forming the mass into a body of the desired shape and provided with continuous channels, drying the body to substantially remove the added water, at least 10% of the added water being removed at a temperature lower than $50^{\circ}C$, and heat-treating the dried body at $800^{\circ}-1000^{\circ}C$ to impart strength and porosity to the body. Preferably the α $-Al_2O_3$ monohydrate is used in an amount of 40-60 w.% with respect to the total weight of the aluminas and the fibres are used in an amount of 1-10 w.% with respect to the total weight of the aluminas.

Condea Petrochemie GmbH, (DE) disclose in Patent (68) a process for producing shaped articles of high pore volume, obtained by extruding aluminium oxyhydrate. In this process alumina hydrate is peptised with an inorganic or organic acid and then diluted with an ammonium solution or with a solution, capable of releasing NH_3, whereafter the obtained mass is extruded, dried and sintered. As aluminium oxyhydrate a boehmite-modification is applied with an Al_2O_3 content of 65-85 w.%, a bulk weight of 500-700 g/l, a specific surface of 140-300 m^2/g (according to B.E.T.) and an average grain size of less than 100 μm. The obtained product is suitable for the manufacture of catalyst carriers.

Patent (245/1) of Sumitomo Aluminium Smelting Co.Ltd, (JP) refers to a hollow catalyst carrier and hollow catalyst made of a transition-alumina (e.g. γ-alumina), which comprises a calcined product having a pipe-like or multi-cell structure and having a very large void ratio (i.e. not less than 3%), a very large specific surface area (i.e. not less than 5 m^2/g), a bulk density of 0.8 to 1.8 g/cm^3, a compressive strength in the extrusion direction of not less than 20 kg/cm^2 and at least one hole in the extrusion direction. This product can be produced by subjecting a powder containing a rehydratable alumina to an extrusion molding, rehydrating the molded product and followed by calcining. The hollow catalyst carrier and hollow catalyst have excellent properties for

carrying the catalytically active components thereon and have an
excellent mechanical strength.

7.2 Zeolites

BASF A.G.(DE) reveal in Patent (24/3) a process for producing nitrogen-
containing, crystalline metal silicates of zeolite structure, from
silicium dioxyde or metal oxides or metahydroxides of Al or B. Crystallisation is effected (with no alkali present) in an aqueous solution of
hexamethylene diamine, applied in an amount of 20-75%, under the self-
pressure of the solution, at 100°-200^oC. The zeolites can be used as
catalysts in the conversion of methanol or dimethylether into unsaturated
carbohydrates, in the oligomerisation of olefines, alkylation of aro-
mates or other conversions of carbohydrates.

Another Patent (24/4) of BASF A.G. refers to the production of C_2-C_5-
olefines on the basis of crude methanol and/or dimethyl ether through
catalytic conversion at raised temperature in the presence of zeolite-
containing catalysts, the zeolite being prepared with the aid of hexamethylene diamine from water glass, without any further addition of a
metal salt.

The zeolite claimed by BASF A.G. in Patent (24/5) is a crystalline iso-
tactic zeolite of the following composition:

$$2/_n R \cdot W_2O_3 \cdot m \, YO_2 \cdot p \, H_2O$$

where R = an organic amine and n = the number of amine groups in the
molecule, W = the elements B, Al, Ga, Fe; Y = the elements Si and Ge.
Such zeolites are obtained from mixtures of SiO_2 and AlOH through hydrothermal crystallisation at 100°-200^oC. The number m depends on the
chain length of amines used in crystallisation.

Yet another Patent (24/6) of BASF A.G. describes the production of crystalline metal silicate zeolites (type ZBM-30), corresponding to the formula:

$$\frac{2R}{n} \cdot W_2O_3 \cdot mYO_2 \cdot pH_2O$$

wherein: R = an organic amine; n = the number of amine groups in the amine molecule; W = the elements B, Al, Fe and Ga; Y = the elements Si and Ge, while p = 0 to 160 and m = a value

$$\geq \frac{96 - \frac{40\,n}{L_R + K}}{\frac{5\,n}{L_R + K}}$$

L_R indicating the length of the amine molecule ($\mathring{A}$) and K = a constant (of about 4). The production process consists of hydrothermal crystallisation.

According to Patent (61/2), granted to the Chevron Research Company (US) the preparation of crystalline aluminosilicate zeolites, exhibiting the X-ray diffraction patterns characteristic of ZSM-5 zeolites, is effected using conventional synthetic procedures but replacing the usual tetrapropylammonium ion source by a source of ethylenediamine, which is relatively inexpensive compared with organic compounds containing tetrapropylammonium ions. Thus ZSM-5 zeolites can be produced by: (a) preparing a mixture of water, a source of ethylene-diamine, and sources of alkali metal oxide, aluminum or gallium oxide, and silicon or germanium oxide; (b) allowing crystals of said zeolite to form in said mixture; and (c) recovering the said crystals of zeolite. The ZSM-5 zeolites produced in this manner can be used in catalysts for hydrocarbon conversion processes.

A crystalline zeolite, designated as CZH-5 has also been developed by Chevron Research Company (61/3), the zeolite being prepared from organic nitrogen containing cations derived from choline. The zeolite, as synthesised and in anhydrous form, has the following composition (expressed as mol ratios of oxides): (0.5 to 1.4)R_2O:(0 to 0.50)M_2O:W_2O_3:xYO_2 where R is an organic nitrogen containing cation derived from choline, x is greater than 5, M is an alkali metal cation, preferably sodium, W is aluminium and/or gallium, preferably aluminium, and Y is silicon and/or germanium, preferably silicon. This zeolite is also useful as a catalyst in hydrocarbon conversion processes.

In Patent (112/2) W.R.Grace & Co. (US) disclose a cyclic method of forming crystalline ZSM-5 zeolite catalyst, which comprises: (a) taking a mother liquor from a previous synthesis and adding the necessary make-up materials such as an oxide of silicon, (b) maintaining the system at a temperature of from about $70^{\circ}C$ to the boiling point of the system and at atmospheric pressure for a time sufficient to permit crystallisation, (c) separating the mother liquor from the desired solid product which is also recovered, and (d) recycling the mother liquor back into step (a).

Imperial Chemical Industries Ltd. (GB) developed several zeolite compositions. Thus, the zeolite claimed in Patent (128/2) is of the EU-2 type and has a molar composition expressed by the formula: 0.5 to 1.5 $R_2O:Y_2O_3$: at least 70 XO_2:0 to 100 H_2O, wherein R is a monovalent cation or $1/_n$ of a cation of valency n, X is silicon and/or germanium, Y is one or more of aluminium, iron, gallium or boron, and H_2O is water of hydration, additional to water notionally present when R is H. The zeolite is prepared from a reaction mixture containing XO_2 (preferably silica), Y_2O_3 (preferably alumina) and an alkylated derivative of a polymethylene diamine, an amine degradation product thereof or a precursor thereof. The zeolite is useful in catalytic processes, especially for the conversion of methanol to hydrocarbons.

The zeolite described in Patent (128/4) of Imperial Chemical Ltd. essentially corresponds to type (128/2), being prepared from a reaction mixture containing XO_2 (preferably silica), Y_2O_3 (preferably alumina), furthermore an optionally substituted quinuclidinium ion.

The EU-1 type catalyst of Patent (128/5) of Imperial Chemical Ltd. is prepared essentially according to (128/2) and (128/4), the reaction mixture from which it is prepared containing in addition to the aforementioned components (silica, alumina) a dicationic alkylated polymethylene diamine. This zeolite is mainly useful in xylene isomerisation.

The zeolite revealed in Patent (128/9) of Imperial Chemical Ltd. essentially corresponds to the aforementioned types, the aqueous reaction mixture, from which a zeolite is obtained, comprising at least one oxide XO_2, at least one oxide Y_2O_3 and at least one tetramethyl ammonium compound, the aqueous mixture displaying the following molar composition:

XO_2/Y_2O_3 5 to 50; free MO_2/XO_2 0.1 to 1.0; Z^-/Y_2O_3 0 to 5000; Q/Y_2O_3 0.1 to 150; H_2O/XO_2 5 to 200, wherein X, Y and M have the given meanings, Z^- is a strong acid radical and $Q = (TMA)_2 + xA$, wherein TMA is a tetramethylammonium compound, A is a trialkylamine or alkonolamine or salt thereof.

The novel zeolites NU-6(1) and NU-6(2) disclosed in Patent (128/6) of Imperial Chemical Ltd. are composed of components, applied in the following proportions: 0.5 to 1.5 R_2O : Y_2O_3 : at least 10 XO_2 : 0 to 2000 H_2O, wherein R is a monovalent cation or 1/n of a cation of valency n, X is silicon, and/or germanium, Y is one or more of aluminium, iron, chromium, vanadium, molybdenum, antimony, arsenic, manganese, gallium or boron, and H_2O is water of hydration additional to water notionally present when R is H. The zeolites are obtained from a reaction mixture containing XO_2 (preferably silica), Y_2O_3 (pref. alumina), furthermore a 4,4'-bipyridyl compound. At a temperature of 200°C or higher Nu-6(1) is converted to Nu-6(2), also a useful catalyst for xylene isomerisation.

The zeolite claimed in Patent (128/10) of Imperial Chemical Industries Ltd. is a member of the ZSM-12 family of zeolites and is prepared from an aqueous reaction mixture containing the oxide XO_2, the oxide Y_2O_3, and a piperazine compound where X is silicon or germanium and Y is aluminium, gallium, boron, iron, chromium, vanadium, molybdenum, arsenic, antimony or manganese.

The zeolite, suitable for catalysing methanol conversion and described in Patent (128/11) of Imperial Chemical Industries Ltd. is produced essentially according to the aforementioned Imperial-Chemical-Patents, the aqueous reaction mixture, the zeolite is made of, containing in addition to XO_2 and Y_2O_3, at least one alkyltrimethylammonium or dialkyl-dimethyl ammonium compounds of which the alkyl groups contain 2 to 10 C atoms.

Yet another synthetic zeolite: "NU-10" is described in Patent (128/12) of Imperial Chemical Industries, which is prepared by the reaction of an aqueous mixture containing XO_2, Y_2O_3 and, optionally at least one organic compound selected from: (a) compounds of the formula: $L^1-(CH_2)_n-L^2$, wherein each of L^1 and L^2, independently, represents a hydroxyl or an optionally substituted amino group and n is an integer from 2 to 20, provided that when both L^1 and L^2 are optionally substituted amino groups, n is an integer from 6 to 20, and (b) heterocyclic bases, the reaction mixture having the molar composition: XO_2/Y_2O_3 = 10 to 10000; M^1OH/XO_2 = 10^{-8} to 1.0; H_2O/XO_2 = 10 to 200; Q/XO_2 = o to 4; M^2Z/XO_2 = o to 4.0, wherein each of M^1 and M^2 represents an alkali metal, ammonium or hydrogen, Q represents the organic compound, X and Y have the meanings given above and Z represents an acid radical. Zeolite Nu-10 is useful as a catalyst for various hydrocarbon conversion reactions, for the conversion of small oxygen-containing organic molecules to hydrocarbons and as an agent for the removal of organic compounds from aqueous effluents.

Patent (180/1) of Mobil Oil Corporation (US) provides a new form of zeolite ZSM-12, containing chromium and/or iron in the as-synthesised form and only a small amount of aluminium, and possessing the x-ray diffraction pattern characteristic of ZSM-12. The proposed zeolite ZSM-12 has a composition according to the formula (in terms of moles of anhydrous oxides):

$$0\text{-}8\ M_{2/n}O : \left[(a)\ Cr_2O_3 + (b)\ Fe_2O_3 + (c)\ Al_2O_3 \right] : 100\ SiO_2$$

wherein M is at least one cation of valance n, a = 0 to 4, b = 0 to 5 and c = 0.001 to 0.5, provided that a and b cannot both be 0 and that, when either a or b is 0, b or a respectively must be greater than c, or according to the formula:

$$0\text{-}4\ R_2O : 0\text{-}4\ M_{2/n}O : \left[(a)\ Cr_2O_3 + (b)\ Fe_2O_3 + (c)\ Al_2O_3 \right] : 100\ SiO_2$$

wherein R_2O is the oxide form of an organic compound containing an element of group 5 - A of the Periodic Table which compound comprises one or more alkyl or aryl groups having from 1 to 7 carbon atoms at least one of which is ethyl. R can be tetraethylammonium and M sodium or potassium.

Patent (180/2) of Mobil Oil Corporation reveals a new and highly siliceous form of zeolite ZSM-11, ways of synthesising it and its use in conversions of organic compounds such as cracking, reforming and alcohol aromatisation. Two species are revealed, one of them containing chromium and/or iron. Synthesis of these species is effected in the absence of added alumina. The products all manifest the x-ray diffraction pattern characteristic of

zeolite ZSM-11. The zeolite's composition is expressed by the formula
(in terms of moles of oxides):

$$(0 - 10)\ M_{2/n}O : (c)\ Al_2O_3 : 100\ SiO_2$$

in which M is at least one cation of valance n and c is measurable but
no greater than 0.5.

The crystalline zeolite according to Patent (180/4) of <u>Mobil Oil Corporation</u>
displays the following composition (expressed in terms of moles of anhydrous
oxides per 100 moles of silica):

$$(W)Q_2O \cdot (X)M_{2/n}O \cdot (y)L_2O_3 \cdot 100\ SiO_2$$

wherein Q_2O is the oxide form of an organic compound containing an element
of Group 5-B of the Periodic Table, this organic compound having at least
one alkyl or aryl group, at least one of which is an ethyl, M is an alkali
metal, $W = < 5$, $X = < 2$, $Y = < 4$, and $L = Al$, Cr, Fe, or La and mixtures
thereof, provided that at least one of Cr, Fe and La is present and Y is
$>$ zero. The new crystal is prepared from a reaction mixture comprising
a source of silica, chromium and/or iron compounds and/or lanthanum oxide,
a source of organic compounds of Group 5-B, having at least one ethyl
group, and water, in the absence of any added alumina in the recipe.

Another zeolite (ZSM-39) described in Patent (180/5) of <u>Mobil Oil Corporation</u>
has a distinctive x-ray powder diffraction pattern, by which it is defined.
It can be prepared from a reaction mixture comprising a source of silica,
a source of organic compounds of Group VB, alkali metal cations and water,
with or without a source of alumina. It usually has a composition, expressed
in oxide mole ratios in the anhydrous form, $(0-2.5)\ M_{2m}O : (0-2.5)Al_2O_3 :$
$(100)\ SiO_2$, in which M is at least one cation of valance n.

Patent (181/2) of <u>Montedison S.p.A. (IT)</u> relates to zeolites having the
formula: $(1.5 \pm 0.6)M_{2/n}O \cdot Al_2O_3 \cdot YSiO_2 \cdot ZH_2O$, wherein Y ranges from
20 to 90, Z ranges from 2 to 12, M is at least one cation and n is the
valence of M. The preparation process comprises the admixing of diethyl-piperidinium hydroxide $[(DEPP)^+OH^-]$ or of a salt thereof with water and
with: (a) at least one sodium compound; (b) at least one compound of
aluminium; (c) at least one compound of silicon; the respective
molar ratios, expressed as ratios of oxides, being comprised in the
following ranges: $SiO_2 : Al_2O_3$ = from 20 to 120; $Na_2O : SiO_2$ =

from 0.07 to 0.50; $(DEPP)_2O : SiO_2$ = from 0.05 to 0.50; $H_2O : Na_2O$ = from 50 to 600, The zeolites obtained according to the invention can be used in converting hydrocarbons by means of reactions usually catalysed by acids, such as cracking, hydrocracking, isomerisation, aromatisation, polymerisation, alkylation, disproportionation, reforming and dealkylation.

Patent (228/3) of Shell Internationale Research Maatschappij refers to a process for the preparation of crystalline iron silicates from an aqueous mixture of the following components: one or more compounds of an alkali metal (M), one or more amines with the general formula $R_1R_2R_3N$, in which R_1 represents an alkyl group and R_2 and R_3 represent an alkyl group or a hydrogen atom, one or more silicon compounds which yield, after drying at $120°C$ and calcining at $500°C$, a product with an SiO_2 content higher than 90w.% and one or more iron compounds, in which mixture the compounds are present in the following molar ratios (with the exception of the amines) expressed in moles of the oxides:

$$M_2O \quad : SiO_2 = 0.01 - 0.35,$$
$$R_1R_2R_3N : SiO_2 = 0.04 - 1.0,$$
$$SiO_2 \quad : Fe_2O_3 > 10, \text{ and}$$
$$H_2O \quad : SiO_2 = 5-65,$$

This mixture is maintained at elevated temperature until the crystalline silicate has been formed, which is subsequently separated from the mother liquor and calcined. The crystalline iron silicates are claimed to have excellent catalystic properties.

CHAPTER 8

MISCELLANEOUS

Alveolated silicate particles can be obtained according to Patent (24/2) of BASF A.G. by preparing droplets of an aqueous solution of an alkaline silicate, which then are solidified in a precipitation bath, transforming them into a gel, that contains a pore-forming agent, whereafter the gel particles thus formed are expanded. The precipitation bath contains mineral and organic gel-forming agents and mineral or organic acids or a combination of these agents, which realise a sudden transformation of the aqueous alkaline silicate solution into a solid gel.

Patent (51/9) of The Carborundum Company relates to preparing a material which mainly contains SiP_2O_7. More particularly, the process consists of mixing together 5 - 95 w% SiP_2O_7 with ZrP_2O_7, which forms the rest of the mixture, that is also added a volatisable organic binder, followed by compressing the mixtures at 28.10^6-140.10^6 Pa, thereby obtaining plates, which then are sintered at a temperature between 1080° and 1190°C, permitting the volatilisation of the binder. At the end, SiP_2O_7 contains practically no $Si_2P_2O_9$.

According to Patent (54/2) of the Centre de Recherches Métallurgiques (BE) the quality of refractory materials (stones, bricks) can be improved by incorporating into the material (between the grains thereof) a finely dispersed material, which reacts more intensively with the metallurgical slag, than the refractory material proper. The refractory material is composed on the basis of magnesium oxide or chrome, iron ore, while the reinforcing agent is incorporated by impregnation and contains Cr_2O_3, Al_2O_3Mg or CaO or ZrO_2.

In Patent (78) of The Dow Chemical Company (US) several methods are set forth for preparing polymetallic spinels by coprecipitating two or more

metal compounds in a proportion to provide a total of eight positive
valences when combined in the oxide form in the spinel crystal lattice.
The methods disclosed require coprecipitation of the metals in the hydroxide form or convertible to the hydroxide-oxide form, calcining the
coprecipitate, and finally sintering the calcined material at about one-
half its melting point or greater, thereby forming a spinel which has a
density of greater than 50 percent of the theoretical density of spinel
crystals. Also disclosed are techniques for preparing less dense spinels,
spinels having more than two metals incorporated into the spinel lattice,
as well as a separate oxide phase associated with the spinel crystallites,
and slipcasting compositions.
More particularly the coprecipitate composed of a layered crystallite
has the following structure:

$$\frac{M^{II} \quad _{d}^{c} Y _{c}^{d}}{}$$
$$\frac{M^{I} \quad _{b}^{a} X _{a}^{b}}{}$$
$$\frac{M^{II} \quad _{d}^{c} Y _{c}^{d}}{}$$

wherein: M^{I} represents one or more metal cations, having valence(s) a,
M^{II} represents one or more metal cations, at least one of which is different
from M^{I}, having valence(s) c different from a; X and Y each represent
one ore more anions having valences b and d, in charge balance with a and
c, respectively, and X and Y are convertible to the oxide on heating;
the molecular ratio of $M^{I}X$ to $M^{II}Y$ being $(1+z)M^{I}X \cdot 2M^{II}Y$ where $z \geqslant$ zero
but less than 3; and sufficient segregate. phases of the formula
$M^{II} \cdot O \cdot Y$ and/or $M^{II}Y$ to provide an overall stoichiometry of $M^{I}x.2M^{II}Y$.

Patent (83/1) of P.Dufour (FR) discloses a process for preparing a vitrifiable composition of a metal silicate or an earth-alkali silicate by
mixing together a salt of a metal or an earth alkali element with an
alkali silicate (in aqueous solution), having an acid content equal to
the maximum amount, which is stoichiometrically necessary for realising
the salification of the alkaline elements of the said alkaline silicate,
which are not salified by the metallic or earth-alkali elements. As
the metal salt, a salt of lead, preferably lead nitrate is recommended.

Another vitrifiable composition of P.Dufour described in Patent (83/2)
is obtained by forming a solution of an alkaline silicate, followed
by a reaction of a soluble salt of a metal (earth alkali metal) modifying agent in the presence of a quantity of free acids, which is sufficient for the salification of the alkaline component. The modifying
agents are non-colouring oxides of Ba, Pb, Ca, Zn, Mg, Cd, Al, etc.

The Institut Khimicheskoi Fiziki Akademii Nauk SSSR (USSR) provide in
Patent (132/1) a mineral refractory material by mixing together an oxide
of one of the metals of group IV, V or VI of the P.S., a metallic or
non-metallic reducing agent or an oxide of a non-metal, whereafter a
small portion of the mixture surface is ignited, thereby starting the
propagation of the combustion zone in the mixture, combustion taking
place under a pressure of the gaseous medium of 1 to 5000 AT (about
1 to 5066 bars). To the initial mixture there can be added nickel,
cobalt, molybdenum (or oxides thereof) in an amount of 5-20 w.%, furthermore manganese and magnesium in an amount of 1-5 w.%.

The Nuclear Energy Research Atomic Energy Council (China/Taiwan)
describe in Patent (135) a method for producing fine uranium powder by
the thermal disintegration of electrolytically obtained uranium amalgam
in vacuum (10^{-4} torr) or in an inert gas atmosphere. The uranium amalgam
may be set into reaction in the presence of (or after treatment with)
methane, at a temperature between $500^\circ C$ and $700^\circ C$. In place of methane
steam or nitrogen may be used as well.

Patent (179) of Mizusawa Kagaku Kogyo K.K. (JP) concerns a process of
producing lead compositions according to the formula:

$$nPbO \cdot PbX_{2/x}$$

wherein: X = a rest of a mineral or an organic acid; x = the valence
of the rest of X and n = a number between 0 and 5. The process consists
of a reaction between lead monoxide (with a vol.weight of 8.3-9.2 g/cm^3),
a mineral acid, an acid oxide and an organic acid in the presence of a gel
of hydroxylamine acid, furthermore a reduction additive.

Patent (242) granted to the Studiengesellschaft Kohle mbH (DE) refers
to alkaline metal complexes according to the structural formulae:

I   II

wherein Me = an alkali metal; x = sulphur or oxygen; a = a whole number
between 3 and 20; L and L' = mono or polyfunctional ethers or amines;
p and q = whole numbers between 0 and 4; R^1, R^2, R^3, R^4 = hydrogen,
radical of alkyl, cycloalkyl, aralkyl or aryl and/or two or more of these
radicals, enclosed in cycloaliphatique or aromatic system. As an alkali
metal lithium or sodium can be used, in finely dispersed form.

Patent (265/2) of Union Carbide Corporation refers to finely divided compositions of oxygen and a metal by contacing a carbohydrate with a metal
composition, followed by their calcination, agglomeration and crushing
into a finely dispersed mass, with a particle size under 1 micron. The
carbohydrate-type component may be paper pulp or cotton; saccharose or
sirup of inverted sugar, while the metal component can be selected from
a great variety of metals, like iron (ferrite), beryllium, magnesium,
calcium, metals of group III/B, V/B, niobium, tantalum, metals of group
VI/B, manganese, cobalt, nickel, copper, zinc, cadmium, aluminium, tin,
lead, bismuth.

In Patent (272/1) of Varta A.G.(DE) a new series of mixed crystals is
described, represented by the general formula:

$$Na_{1+x}M_{2-1/3x+y}Si_xZ_{3-x}O_{12-2/3x+2y}$$

wherein: M = Zr, Ti, Hf or their mixture; Z = P, Sb, Bi, V, Nb, Ta or
their mixture; x = 0.01-3 and 4 = 0-0.5. In case of x-values under 1.5
the crystals have a rhomboidal form, while in case of x-values over 1.5,
single-phase monocline crystals are forming, which, due to their high
crystallisation degree, can achieve extremely high densities during
pelleting, and which with regard to their stability against Na and to
their great effectiveness, are equal to Na - β - aluminium oxide
(Na_2O - $11Al_2O_3$) compositions. In general, the crystals are resistant

to humidity and, being effective Na^+ ion conductors, can be used as
solid electrolyses in galvanic cells.

Patent (286) of Yi Hung Fang, A. (US) relates to a process for reclaiming
phosphate products from aqueous acid residual solutions, which form
during the immersion treatment of polishing aluminium, the process comprising a reaction between the aqueous acid residual solution and an
alkaline substance, like soda, a composition of soda and sodium carbonate, thereby transforming aluminium phosphate and sodium aluminate
and sodium triphosphate, which are present in salt form in the aqueous
reaction solution and from which most of the phosphate content can be
reclaimed by crystallisation.

An improved method of preparing silicon carbides consists, according to
Patent (79/4) of The Dow Corning Corporation (US) of forming a desired
shape from a polysilane of the average formula: $[(CH_3)_2Si]$ $[(CH_3Si]$.
The polysilane contains from 0 to 80 mole percent $(CH_3)_2Si$ units and
from 40 to 100 mole percent CH_3Si units. The remaining bonds on the
silicon are attached to another silicon atom or to a halogen atom in
such a manner that the average ratio of halogen to silicon in the polysilane is from 0.3 :1 to 1 : 1. The polysilane has a melt viscosity at
$150^{\circ}C$ of from 0.005 to 500 Pa.s and an intrinsic viscosity in toluene
of from 0.0001 to 0.1. The shaped polysilane is heated in an inert
atmosphere or in a vacuum to an elevated temperature until the polysilane
is converted to silicon carbide.

PART II

INTERMEDIATE PRODUCTS

AND

GREEN BODIES

CHAPTER 9

Ceramic and hard materials, intermediates, based on aluminium and aluminium oxides

9.1 Ceramic and hard materials, intermediates based on alumina

Patent (11/2) granted to <u>Aluminium Pechiney (FR)</u> reveals a process for preparing agglomerates of aluminium oxide, displaying high mechanical strength and controllable granulometry. The agglomerates are obtained by compacting an intermediary product, obtained from the incomplete decomposition of hydrated aluminium nitrate, containing 0.5 to 15 w.% N_2O_5, followed by granulation of the compacted material and heat treatment of the granules obtained. From this material various products can be made like balls, cylinders, small plates, etc.

Another Patent (11/3) of <u>Aluminium Pechiney</u> refers to aluminium oxide agglomerates of high mechanical strength, obtained by compacting an intermediary product, derived from the incomplete decomposition of hexa-hydrated aluminium chloride, containing 0.8 to 15% chlorine. The inter-mediary product is compacted under a pressure of 2000 to 10,000 kg/cm². The agglomerates display after compacting, but before heat treatment an apparent specific weight of 1 to 1.5 g/cm³. Heat treatment takes place between 600° and 1500°C. Similarly to (11/2) balls, cylinders, small plates, pills can be obtained from the described material.

Patent (14/1) of <u>American Cyanamid Company (US)</u> provides extrusion additives, suitable for promoting the extrusion of an aluminium composition, which contains a considerable amount of rehydratable aluminium oxide and water. Such additives may contain precipitated aluminium oxide powder, concentrated bauxite and aluminium oxide flour, furthermore an agent for

intensifying extrusion, consisting of methyl cellulose, tall oil, surface-active and flocculating substances.

In another Patent (14/2) of American Cyanamid Company a process is described,
producing an alumina support of low density, composed of 60-90 w.% of
rehydratable alumina; 10-40 w.% of microcrystalline cellulose and a sufficient amount of water for obtaining a solid matter content of 45-70%
in the mixture. The mixture is extruded and the extrudate is hardened
and calcined at 926-1093°C.

Patent (23/1) of The Babcock & Wilson Company (US) concerns an alumina
based, phosphate-bondable, dry refractory mix composition, exhibiting
an extended shelf life of up to eighteen months and longer, with the
following composition (on a weight percentage basis) 40 - 70% of an
inert refractory aggregate, selected from tabular alumina, calcined
bauxite, kaolin calcine and synthetic mullite; 15 - 35% of an aluminous
material selected from calcined alumina, calcined bauxite or kaolin
calcine; 2 - 10% hydrated alumina; 1 - 10% calcium aluminate cement;
and an acid ingredient in an amount providing P_2O_5 equivalent to that
provided by 3 to 19 weight percent of an 85 percent concentration of
phosphoric acid. The mix is prepared by feeding the dry ingredients
into a mixer, mixing the dry ingredients to a homogenous state, then
adding a liquid acid ingredient to the mixture of dry ingredients, and
subjecting the dry and acid ingredient mix to a high intensity mixing
action for a predetermined period of time. The product is substantially
dry to the touch.

According to Patent (51/19) of The Carborundum Company an abrasive composition is obtained by a smelt mixture of bauxite and zirconium, the
smelt mixture being solidified in about 3 minutes after the start of
the cooling period in homogeneous state. The composition contains
about 35-50 w.% zirconium, 1.1 - 2.5 w.% silicium oxide, furthermore
about 2.5 w.% titanium dioxide.

"Chemokomplex" Vegyipari Gép. Es Berendezés Export-Import Vállalat
(Hungary) present in Patent (59) a process for producing sintered ceramic
products based on rocks, containing 40-85% SiO_2 and 5-20% Al_2O_3, the rocks
being either underground or surface formations or the debris of quarries

and having a fusion point under 1450°C. The initial material is crushed
into particles of smaller than 20 mm, whereafter the material assorted
according to particle size, is mixed separately with clay, kaolin or
bituminous slate (slags), the mixture then being crushed into fractions
under 0.25 mm, granulated or formed into a paste (with water) and
placed in forms for firing.

Patent (70/1) of <u>Corning Glass Works (US)</u> refers to the production of
aluminous keatite (a high-pressure synthetic form of silica, not yet
found in nature) ceramics, by contacting a ceramic article having a
mole ratio of Al_2O_3 to modifying oxides selected from Li_2O, Na_2O, K_2O,
ZnO, MgO, CaO, BaO and SrO not exceeding 1 : 1 and having a principal
crystal phase composed of a beta-spodumene solid solution with an
acid in order to replace at least some of the lithium ions therein
with hydrogen and subsequently heating the resulting article to provide
aluminous keatite by the removal of at least some of the water of crystallisation therefrom.

<u>Corning Glass Works</u> also present in Patent (70/2) a three-phase aluminium-
zirconium-silicium oxide ("AZS") system with a closely controlled fraction
composition of the components, according to which the system contains:
<u>a</u>) 25-60% poured refractory coarse AZS grains, 89% of which has a particle
size of 0.833-3.327 mm, <u>b</u>) 0-38% medium AZS grains, about 73% of which
has a particle size of 0.417-1.65 mm and <u>c</u>) 30-50% of at least one of
the components: (i) fine AZS grains, (ii) aluminium oxide (with 98.5%
Al_2O_3 and max. 0.5% Na_2O); or chrome oxide (with min. 95% Cr_2O_3) or
ultrafine AZS grains (with 96% of size 0.147 mm and at least 20% of size
0.043 mm).

According to Patent (77/1) of <u>Didier Werke A.G.</u> refractory concrete is
produced from used, broken up tank blooks of $ZrO_3-Al_2O_3-SiO_2$, the particles
of the broken elements having a size under 6.3 mm and form 55-90 w.%
of the mixture. More particularly the composition consists of 10-13 w.%
ZrO_2; 50-70 w.% Al_2O_3; 5-15 w.% SiO_2; 3-10 w.% CaO and 0-4 w.% other
oxides. The end product (concrete) is suitable for lining glass smelt
ovens.

Feldmühle A.G.(DE) provide in Patent (94/3) shaped ceramic bodies on the basis of non-metallic hard materials (metal carbides, metal nitrides, metal borides and oxides of aluminium and zirconium) the bodies also containing one or more eutectic structural components of beryllium oxide, zirconium oxide and/or hafnium oxide. The shaped bodies can be used in producing turbine blades, friction discs, valve packing discs or cutting plates. The components obtained by eutectic reaction display a high degree of fineness. Due to the transformation from the metastable tetragonal into the monocline modification under mechanical load, zirconium oxide and/or hafnium oxide are capable of high energy absorption.

Patent (96/9) of Ford France S.A. (FR) concerns a method of sintering a shaped green, beta-type alumina body by inserting the body into an open chamber, prepared by exposing the interior surface of a container consisting, essentially, of at least about 50 weight percent of alpha-alumina and a remainder of other refractory material, to a sodium oxide or sodium oxide producing environment; sealing the chamber and heating the chamber with the shaped body encapsulated therein to a temperature and for a time necessary to sinter the body to the desired density. The sodium oxide containing compound comprises sodium aluminate, sodium acetate, sodium carbonate or Na_2O . XAl_2O_3, wherein X = less than or equal to 11.

Patent (98/5) of Ford Werke A.G. (DE) describes a process for producing anisotropic, polycrystalline sinter ceramics on the basis of cordierite, with less than 0.5 w.% iron-titanate, which is mixed to the pulverulent glass mass as crystal core forming agent. The product is obtained by crushing a glass mass on the basis $2MgO$. $2Al_2O_3$. $5SiO_2$ into a fine powder, the powder then being mixed in a ball mill with 0.5 w.% irontitanate, the formed blend then being sintered at $1095^\circ C$. Prior to sintering 0.5w.% zirconium oxide and 0.2 w.% magnesium fluoride are added to the blend.

Patent (104/22) of General Electric Company relates to alumina-based ceramics for core materials having a good leachability and being non-reactive with the alloy up to $1800^\circ C$. The ceramic material, before firing, consists of at least an alumina-based compound material having the general formula: MO . NaL_2O_3, wherein M is Na, Ca, Sr or Ba, N is from 9 to 11 when M is Na_2 and N is 6 when M is Ca, Sr or Ba, and after firing having a density of from 40 percent to 75 percent of theoretical and an interconnected network defining a plurality of interstices in which undoped alumina grains are found. The ceramic material can be used in making cores for directionally solidified eutectic and superalloy material.

General Refractories Company (US) presents in Patent (105/3) a refractory material on the basis of aluminium oxide or aluminium silicate, suitable for coating aluminium smelting ovens, the said refractory material containing 0.5 - 30 w.% barium sulphate, furthermore a binder on the basis of a phosphate or calcium aluminate. The refractory material can be used in form of bricks, packing material, plastic mixture, mortar and displays a high resistance against the penetration of metals and against the adherence of slag.

In Patent (106) of General Motors Corporation (US) a refractory product of sintered ceramic material is disclosed, composed of magnesium, aluminium, crystalline silica (with cordierite as the main component). The composition displays a high resistance against thermal shocks and a low dilatation coefficient. The smaller crystalline phase of the material consists of mullite. The composition contains $Na_2O + K_2O$ in an amount of not more than 0.14% of the material's weight.

Patent (109) of Globe Union Inc. (US) relates to ceramic bodies of open porosity (25-30%), formed of a particulate material by fusion, like a mineral material of high boiling point (with particle sizes under 600 microns). The mineral material may contain silicon oxide, aluminium oxide, a small amount of glass (smelting point at 540-1370°C) for bonding the mineral material.

Patent (112/1) of W.R.Grace & Co. concerns the production of shaped bodies of aluminium oxide of high density, stability, large surface and low pore volume. The production process of these bodies includes the formation α -aluminium-oxidemonohydrate (particle size: 10-150 /um), to which water is added in order to obtain an extrudable paste, which then is extruded, the extrudate being aged at ambient temperature during 8 to 8 hours at 30-50°C. Drying takes place in an oven, in a coerced air current at 65-132°C.

According to Patent (114) of Groupement Atomique Alsacienne Atlantique (FR) carbide-base sintered ceramic material can be produced by mixing fine powders, one of which consists of a carbide of a IV, V and VI group element of the P.S.; blending the obtained mixture with the fluoride of an element of the group I/a of the P.S. at a 0.1-6 w.% ratio, adding an organic gel (gum arabic); feeding the mixture into a mould and sintering (without charge) at a temperature between 900° and 1800°C. The product displays controlled porosity and is resistant to corrosive metal smelts (of aluminium and steel).

Imperial Chemical Industries Ltd. claim in Patent (128/3) a felt, which consists of polycrystalline synthetic fibres of refractory metals (aluminium, zirconium) and an inorganic binder. The fibres are obtained by spinning from a composition, which contains aqueous oxychloride, basic acetate, basic formiate or nitrate of aluminium and/or zirconium and a smaller amount of an organic, water-soluble polymer. The inorganic binder may be composed of silicium, zirconium, aluminium or fibrous boehmite.

Kaiser Aluminium & Chemical Corporation (US) developed in Patent (145/1) a process for producing a refractory material on the basis of synthetic mullite of increased purity, which contains a finely dispersed hydrated aluminous substance, that displays after calcination an Al_2O_3 content of at least 95% and a finely dispersed aluminium silicate, which contains after calcination at least 95% $Al_2O_3+SiO_2$, the mixture of these components being calcined at 750°-1150°C. The fraction sizes of Al_2O_3 vary between 43 and 10 microns, while those of aluminium silicate vary between 43 and 5 microns.

Patent (147/2) of Kalinin, V.P., et al.(USSR) refers to high temperature resistant ceramic products, obtained by shaping a body from an initial material, consisting of 65-72 w.% Al_2O_3 and 28-35 w.% MgO, the obtained green body being dried by heating to 400°-450°C (at a rate of 30-60°C/min), in air and to 950°-1050°C (at a rate of 30°-100°C) in vacuum. Thereafter follows sintering in a noble gas atmosphere at 1800°-1850°C (at a rate of 100°C/min), and holding the product at this temperature for separating the used heat carrier from it, whereafter the product is again sintered at this temperature.

Another Patent (147/3) of V.P.Kalinin, et al. concerns the production of
refractory ceramic materials on the basis of aluminium oxide and magnesium oxide, these substances being fed into a mould, in the centre of which
there is placed a heating element, the mould content being dried by raising
the temperature of the said heating element at a rate of $30-60^{\circ}$C/min. to
$400^{\circ}-450^{\circ}$C, then at a rate of $30-60^{\circ}$C to $950^{\circ}-1050^{\circ}$C, under vacuum.
The product is sintered in an inert atmosphere by heating the heating ele-
ment at a 100°C/min. rate to $1800-1850^{\circ}$C. Thereafter the product is maintained at $1800^{\circ}-1850^{\circ}$C for a time necessary to realise the free separation
of the heating element from the product.

Patent (152) granted to H.Kiefer and J.M.Allibert (FR) describes a porcelain, obtained by burning a mass, composed of aluminium oxide, quartz,
alkali-earth oxides or compositions, which during burning release the
aforementioned oxides. More particularly, the porcelain contains 40-80 w.%
SiO_2, 20-50 w.% Al_2O_3, 1-15 w.% earth-alkali oxides and alkali oxides,
The mass (in dry state) is shaped into the required forms, shaping and
burning being carried out simultaneously.

In Patent (154/2) Kombinat VEB Keramische Werke Hermsdorf (GDR) claim
the production of non-porous ceramic bodies, on the basis of corundum
and glass, the material the bodies are made of being sintered at a temperature under 1050°C and displaying the following granulometric composition: the fraction under 20 microns in more than 85 w.%; the fraction
under 2 microns in more than 3 w.%, while 64-40 w.% of the composition
consists of crushed glass of earth-alkali aluminium silicates (containing SiO_2, Al_2O_3, B_2O_3, CaO, MgO, BaO, ZnO, Fe_2O_5, F).

Patent (159/1) of Fried.Krupp GmbH (DE) refers to anticorrosive, protective layers for shaped bodies, mainly of a hard metal, whereby at
least one of the protective layers is composed of a ceramic matrix, into
which another material (ZrO_2) is incorporated. The ceramic material
(of Al_2O_3) and the material inserted therein (8-25 vol.% ZrO_2) display
different thermal expansion coefficients (that one of the inserted
material being the lower one) the said layer being traversed by micro-
fissures.

A refractory aluminum oxide-chrome composition can be obtained by a

process described in Patent (164/2) of <u>Kyushu Refractories Co.Ltd.(JP)</u>
containing 30-80 w.% aluminium oxide, silicium oxide, furthermore one of
the following substances: synthetic mullite, sillimanite, cyanite,
andalusite, diaspore, bauxite or the like. The chrome-containing component may comprise chrome ore, chrome-magnesium oxide-klinker or chrome-
magnesium oxide brick particles.

<u>Lafarge S.A. (FR)</u> disclose in Patent (166/1) a refractory material, which
contains aluminium oxide and a small amount of silicium oxide. The green
body of the material is composed of 15-50 w.% of a mixture, which on the
one hand contains crushed aluminous cement (specific surface: 3000-6000
cm^2/g), furthermore the following components: 15-45% CaO; 38-82% Al_2O_3;
0-10% SiO_2; 0-15% MgO; 0-20% iron oxide (expressed as Fe_2O_3), and on
the other hand an aggregate or granulate, consisting of dry hydrated
aluminium oxide and corundum.

Another Patent (166/2) of <u>Lafarge S.A.</u> relates to aluminium oxide-
silicium oxide refractory materials as well, the green body of which
contains aluminous cement, furthermore the components according to
Patent (166/1) and in addition dry hydrated aluminium oxide, corundum
and an alkali phosphate, the latter in such an amount that the mixture
contains 1-15% P_2O_5. The weight ratio between aluminous cement and the
refractory mixture is established between 15 and 50%.

<u>Max Planck Gesellschaft zur Förderung der Wissenschaften e.V. (DE)</u> claim
in Patents (176/2 and 176/5) a ceramic matrix with a ceramic inserting
material dispersed therein for shaped ceramic articles of high resistance to breaking. The ceramic insertion material displays (at the firing
temperature of the shaped body and at room temperature) different enantiotropic modifications of various densities. The shaped ceramic body is
traversed by densely arranged, fine microfissures and it also comprises
a supplementary inserted phase, consisting of a ceramic matrix and a
substance dispersed therein, the amount of which differs from the amount
of the insertion material contained in the initial composition. The
ceramic matrix consists of Al_2O_3 and the inserted phase of unstabilised
ZrO_2 particles.

Another Patent (176/7) of the <u>Max Planck Gesellschaft zur Förderung der</u>

Wissenschaften e.V. refers to the production of ceramic bodies similar to those described in Patents (176/2) and (176/5), the matrix of which consists of Al_2O_3, SiC or Si_3N_4, in which ZrO_2 or HfO_2 particles are dispersed in unstabilised state.

Patent (178/2) of <u>Minnesota Mining and Manufacturing Company (US)</u> concerns a method of making synthetic aluminium oxide-bases abrasive mineral by

(1) providing, in a liquid medium, a substantially calcium ion- and alkali metal ion-free homogeneous mixture of an alumina source compound and at least one precursor of at least one metal oxide selected from the group consisting of cobald oxide, hafnium, magnesium, nickel oxide, and zirconium to provide, on a fired solids basis, a mineral containing one of the following:

   (a) at least 10 vol.% of zirconium, hafnium or a combination of zirconium and hafnium;

   (b) at least 1 vol.% of spinel, derived from the reaction of alumina with at least one of said metal oxides; or

   (c) at least 1 vol.% of at least one spinel as defined in (b) and 1-45 vol.% of a metal oxide selected from the group consisting of zirconium, hafnium and a combination thereof;

(2) gelling the resultant mixture;

(3) drying the gelled mixture to form a solid material; and

(4) firing the solid material in a non-reducing atmosphere to at least $1250^\circ C$, at a controlled rate to remove volatile materials and non-destructively convert the solid material to a dense but non-fused α -aluminium oxide-based mineral.

According to Patent (181/1) of <u>Montedison S.p.A.</u> porous tubes can be produced from aluminium oxide or other oxides of high purity, the production process comprising: granulating aluminium oxide powder by a dry or humid treatment, thereby obtaining granules with 40% density of the nominal density value. The granules may be mixed with a thickening agent and/or binders. The granules then are processed into tubular form, in an isostatic press, the shaped product being sintered in the atmosphere of a combustion gas at a gradually increasing temperature $(1550^\circ-1780^\circ)$ with a 1-hour holding time at the maximum temperature.

Patent (190/8) of NGK Insulators Ltd. (JP) relates to low-expansion
ceramic material, the chemical composition of which consists of from
1.2 to 20 w.% of magnesia, from 6.5 to 68 w.% alumina, from 19 to 80 w.%
of titanium (calculated as titanium dioxide), from 1 to 20 w.% of
silica, and from 0.5 to 20 w.% of iron (calculated as ferric oxide);
the major component of the crystalline phase of the material being a
magnesium oxide/aluminium oxide/titanium dioxide/silicon oxide/iron
oxide solid solution; and the material having a coefficient of thermal expansion of not more than 20×10^{-7} $(1/^{\circ}C)$ in the temperature
range $25^{\circ}C$ to $800^{\circ}C$, a four-point flexural strength of not less than
50 kg/cm^2 at room temperature, and a melting point of not less than
$1500^{\circ}C$. The material is prepared by forming a batch of magnesium,
alumina, titanium oxide, silica and iron oxide (as such or in the form
of precursors therefor capable of being converted thereto on firing);
plasticising the batch if necessary and shaping the batch thus shaped;
and firing the shaped body at $1300^{\circ}C$ at $1700^{\circ}C$. The ceramic material
is suitably in the form of a honeycomb structure.

In Patent (195/3) Nippon Gaishi K.K. (JP) reveal   a polycrystalline
spinel sintered body consisting mainly of Al_2O_3 and MgO in a molar ratio
of Al_2O_3/MgO of from 0.52/048 to 0.70/Q30 and containing 0.001-0.1 w.%
of LiF, which can be obtained by calcining a powdery mixture of Al_2O_3
and MgO in a specifically limited molar ratio, moulding the calcined
mixture into a shaped article together with a specifically limited
amount of LiF, and firing the shaped article under specifically limited
temperature conditions.

Patent (197) of Nippon Kouatso Electric Corp. (JP) refers to porcelain,
calcined at a low temperature, the porcelain being composed of a glassy
substance, aluminium oxide, a clayey component, a binder and (as the case
may be) a colouring agent. The binder can be of phosphoric acid or
phosphate, while the colouring agent could be a metal oxide. The glassy
substance consists of shavings,chips of glass.

Patent (204/2) of Norton Company refers to a shaped, alumina body, having
a surface area in excess of 60 square meters per gram and having at least
0.115 cubic centimeter per gram of pores larger than 350 ångstroms diameter, at least 0.1 cubic centimeter per gram of pores larger than 1000

ångstroms diameter, the porosity being exclusive of any pore volume produced by particulate burn-out material, the body consisting of over 99 w.% of alumina, exclusive of combined or adsorbed water and exclusive of added catalytic metal or metal oxides, the alumina being predominantly in a form selected from the group consisting of gamma alumina, boehmite, delta alumina, theta alumina and mixtures thereof.

Patent (205/2) of Novatome Industries (FR) relates to bi-phase ceramic articles composed of silicium oxide, aluminium oxide and magnesium and is prepared by introducing into an aggregate of aluminium oxide, silico-aluminium oxide or magnesium a gel of aluminium-silicium oxide (with 50-75% aluminium oxide and 50-25% silicium oxide), adding to the aggregate the gel in an amount of 2-10% (in dry state), followed by dehydration and the mechanical dissociation of the obtained mixture, which then is formed into a paste, of which shaped articles are made and sintered at about 1400°C.

Patent (226) of Schweizerische Aluminium A.G.(CH) describes a porous ceramic filter of open foam structure for molten metals. More particularly, the filter is composed of a ceramic material, which contains 55-70 w.% Al_2O_3; 2-10 w.% reactive aluminium oxide (with micron-sized particles); 1-5 w.% montmorillonite and 1-10 w.% fibres of aluminium silicate. The ceramic material is made into a paste with which an organic foam element (with open pore structure) is impregnated. Thereafter, the paste content is squeezed out to such an extent, that the foam structure will be completely coated with it. This combination is then heat-treated, thereby burning out the organic foam and hardening the ceramic material.

Patent (237/2) of Société Européenne des Produits Réfractaires provides lining elements for gravity-discharge ovens (for steel production), the elements being composed of molten refractory material of the following chemical components: 10-28w% ZrO_2; 3-12 w.% SiO_2; 60-80 w.% Al_2O_3; 0.3-1.5 w.% Na_2O; 5 w.% of TiO_2, CaO, MgO. The crystallographic composition is the following: 60-80% corundum; 10-28% zirconium; 5-19% vitrified phase; corundum + zirconium + vitrified phase, composing 99% of the composition, while the amount zirconium + 2.5 (vitrified phase) corresponds to 33-57.5%.

Patent (247) of <u>Taiko Rozai Co.Ltd. (JP)</u> relates to monolithic refractory compositions for lining various types of smelting ovens or metal
smelt containers, the compositions containing 3-12 w.% refractory clay;
0.1-5 w.% aluminium metal powder, a flocculating agent, (0.01-1.0
w.%) a coagulating agent (1-8 w.%), a reaction inhibitor of the aluminium
powder (ammonium borate), the rest of the composition consisting of particulate refractory aggregates.

Patent (260) of <u>Treibacher Chemische Werke A.G.(Austria)</u> refers to a
process for producing grinding materials on AlO-basis,also in combination
with, for example ZrO, by smelting the oxides, followed by a quick cooling of the smelt, by pouring the oxide smelt into a smelt of a salt or
salt mixture (e.g. calcium chloride or a mix of calcium chloride and
sodium chloride). The Al_2O_3 or Al_2O_3 + ZrO_2 mix contains impurities of
chrome, iron, silicium, titanium, vanadium, calcium, magnesium, and/or
rare earth metals. Heating temperature is 1350°-$1550^\circ C$, while cooling
takes place to a degree 50°-$350^\circ C$ under the solidification point of
the oxide mix.

Patent (267/18) of <u>UKAEA</u> provides a method for preparing sintered
materials, comprising a ceramic matrix (e.g. of Al_2O_3) and, distributed
therein, a refractory material in a metastable, high temperature enantiotropic form of higher density than its room temperature enantiotropic
form such as ZrO_2,which has a metastable tetragonal form and a room
temperature monoclinic form, from a mixture where the refractory material and optionally the ceramic material are in the form of a sol. The
mixture is dried and subsequently sintered to give a product where the
refractory material is distributed very uniformly and in its metastable
form. Improved properties have been obtained by including a stabilising
agent (e.g. Y_2O_3) in the final products.

The <u>University of Guyana (Guyana)</u> developed a method (270) for the manufacture of sodium chloride or brine and/or caustic soda, alumina plant
red mud, a waste product in the manufacture of alumina or metal grade
bauxite. The method comprises: heating alumina plant red mud in the
presence of hydrochloric acid, the hydrochloric acid being added to attain a pH of 5.5 to 6, and separating the solids from the liquid phase.
The solid sodium chloride is recovered from the brine by evaporation

thereof. The treatment of alumina plant red mud is carried out at a temperature of from 70 to 80°C, its drying taking place at a temperature of about 115°C. From the material thus obtained high-loadbearing bricks can be produced by the addition of kaolinitic clay thereto and by firing at 1090-1150°C. Such bricks have a compressive strength of 7×10^7 Nm^{-2} (approx. 10,000 psi) and an average water absorption less than 1.4 w.%. The invention solves storage and pollution problems.

Patent (276/3) of Volkswagenwerk A.G. (DE) relates to refractory heat insulating ceramic materials, composed of a mixture of 30-60 vol.% of powdery aluminium oxide (Al_2O_3) and 70-40 vol.% of aluminium silicate fibres. The mixture also contains zirconium silicate ($ZrSiO_4$) in a proportion of 5-20 w.%, furthermore catalytic additives, selected from the group, that contains: titanium, vanadium, chrome, manganese, iron, cobalt, nickel, copper or oxides or other compositions thereof. The additive may also be a rare-earth element or a noble metal (platinum).

Patent (277) of Vostochny Nauchno-Issledovatelsky I Proektny Institut Ogneupornoi (USSR) provides a process for producing pyrometric refractory components, wherein finely dispersed ingredients are mixed in the following ratio, by w.%: alumina 65 to 83; titanium dioxide 6 to 10; zirconium dioxide 11 to 25. A plasticising additive is introduced into the resulting mixture, the components are moulded and subjected to initial firing at a temperature of 1200 to 1300°C, aimed at complete removal of the plasticising additive and partial strengthening of the components, then to final firing at a temperature of 1650 to 1720°C, aimed at imparting the required strength thereto.

9.2 Alpha- and beta-alumina-based intermediates

Patent (43) of British Railway Board (GB) relates to beta-alumina polycrystalline ceramics, suitable for use as an electrolyte in apparatus of the kind involving the transport of sodium ions by diffusion through a solid electrolyte, for example in electric cells or batteries of the sodium-sulphur type. Experiments proved that the strength and hence the performance and life of the beta-alumina in sodium sulphur cells is connected with the crystal size of the fired beta-alumina, the strength increasing with a consistent small crystal size and that crystal size is connected with the mode of firing used to produce the beta-alumina. Thus, the patent reveals a method of sintering beta-alumina, according to a firing schedule, comprising a temperature rise to about $1500^{\circ}C$, then cooling, then a reheat to about $1600^{\circ}C$ to complete sintering and then cooling to below $1500^{\circ}C$. The temperature rise from about $1200^{\circ}C$ takes place at an average rate of 100° to $150^{\circ}C$ per hour.

Patent (66/8) of Compagnie d'Électricité (FR) concerns the preparation of alkaline beta-alumina articles,(for example with sodium), the sintering process being carried out in a reactor, in an alkali metal-rich atmosphere at a temperature between 1600° and $1700^{\circ}C$ during a period of time varying between 30 minutes and 4 hours. The sintered mass is left to cool to ambient temperature. The reactor is made of a refractory material which contains alpha-alumina carborundum and zirconium.

Patent (119/2) granted to K.Hart (DE) reveals the production of shaped elements of polycrystalline alpha-aluminium oxide containing beta-aluminium oxide, wherein the lamellar crystals of the beta-phase are approx. parallel to one another. It is also possible to produce membrane-shaped elements, wherein the ion-conducting planes of the beta-phase crystallite are approx. perpendicularly oriented to the membrane plane. The shaped elements of sinter corundum are exposed to a sodium oxide vapour pressure at $1100-1400^{\circ}C$.

Patent (63/3) of Chloride Silent Power Ltd. (GB)   refers to the manufacture of beta-alumina solid ceramic articles, containing a doping addition of magnesium oxide and lithium oxide, comprising the steps of wet-milling a precursor material or emulsification of an already powdered precursor material to form an aqueous slurry, spray-drying the slurry to produce a powder, forming the powder into green shape and sintering the shaped product. The precursor material contains at least some of the magnesium oxide in the form of magnesium aluminate. By using spray drying it is possible to produce spherical aggregates of beta-alumina precursor powder, having excellent flow properties.

Impervious,polycrystalline,cationically conductive beta-alumina is obtained according to Patents (63/7) and (63/10) of <u>Chloride Silent Power Ltd.</u> consisting of densifying by sintering a compact of beta-alumina particles or a compact of particles that react together on heating to form beta-alumina in the course of a succession of cycles of heating, followed by cooling, so that not more than 95% of the overall linear shrinkage takes place during any one cycle of heating followed by cooling. The novel sintering procedure takes place in a closed crucible according to a firing schedule, comprising: a temperature rise to a first peak temperature between 1450°C and 1600°C, then cooling, then a reheat to a second peak temperature which is at least 10°C higher than the first, but which does not exceed 1900°C, to complete sintering, then cooling, the rate of heating to the first peak temperature and of the subsequent cooling being such that between 5% and 95% of the overall linear shrinkage occurs in this cycle.

Another Patent (63/11) of <u>Chloride Silent Power Ltd.</u> refers to the production of beta-aluminium ceramic products, obtained by compressing a finely divided initial material, followed by heating to form a homogeneous green body and by sintering in order to obtain an impermeable, polycrystalline ceramic material. The sintered material is exposed to isostatic pressure over 34.5 MPa at 1200°-1550°C, whereafter the material is cooled under pressure to a temperature under 1200°C, after which pressure is suppressed.

In a process, described in Patent (66/1) of <u>S.A. Compagnie Générale d'Électricité (FR)</u> beta alkaline alumina parts are sintered in a sintering chamber, which is at least partially made of a refractory material, that includes a mixture of three ingredients, namely beta alumina fire clay obtained by melting and crushing into grains with a grain size of about 0.5 mm, beta alumina cement or binding agent obtained from an alpha alumina and sodium carbonate mixture by reaction in the solid state and at a temperature of about 1200°C, such a cement having a grain size of about 10 microns and lastly a sodium salt, the respective proportions by weight of these three ingredients causing that during sintering the chamber generates an atmosphere rich in sodium in the immediate vicinity of the parts. The proportions of the three ingredients are: fire clay 75%, cement 20% and sodium salt 5%.

Compagnie Générale d'Électricité also present in Patent (66/4) a process
for producing alkali metal containing beta-aluminium oxide by preparing
an intimate mixture of aluminium oxide powder and the powder of an alkali metal carbonate (e.g. sodium carbonate), this mixture being heated in
an open crucible and then cooled, whereafter the powdery mass is shaped
into an article of required form and sintered. After shaping, the surface of the article is coated with an alkali metal composition under
normal atmosphere. Coating can be effected by immersion, spraying, etc.
The coating solution contains the alkali metal composition in an amount
of 30-350 g/l.

According to Patent (66/6) of Compagnie Générale d'Électricité alkaline
beta-alumina (without the allotropical β " variant) is prepared from
an intimate mixture of alumina powder and at least two compositions of
the same alkali metal, the first composition being selected from the
group of carbonates and aluminates, while the second composition contains
at least a fluoride (e.g. a double fluoride of aluminium). The mixture
is heat treated at $1000^{\circ}-1550^{\circ}C$ in an open atmosphere.

Another Patent (66/9) of Compagnie Générale d'Électricité concerns a
process for joining elements, composed of alkaline beta-aluminium oxide
and alpha-aluminium oxide resp., by heating the alpha-aluminium oxide
element to $1400^{\circ}-1600^{\circ}C$ and holding it at this temperature for 10-20
hours, while it is embedded in a powdery mass of a carbonate or aluminate, followed by cooling to ambient temperature. On the other hand,
there is made a blank from the alkaline beta-aluminium oxide component,
whereafter both elements are placed in a sintering oven, in which an
alkali-metal-rich atmosphere is created and which is heated to $1600^{\circ}-$
$1700^{\circ}C$ during a period of 30 minutes to 4 hours.

According to Patent (66/10) of Compagnie Générale d'Électricité beta-
aluminium (of the allotropic β " variant) articles are prepared by making
an intimate mixture of alumina powder and an alkali carbonate (e.g. of
sodium) and after the usual treatment phases, by forming a sediment of
the said alkali metal on the mass, placed in a mould. In the mixing
phase there is introduced into the mass an oxide of lithium and an oxide
of magnesium at a rate of 0.05-1 mole per 1 mole of alkali oxide.
Sintering is carried out at $1500^{\circ}-1700^{\circ}C$.

Patent (66/12) of <u>Compagnie Générale d'Électricité</u> reveals the preparation of articles of alkaline beta-alumina (with sodium, for example with
an Al_2O_3/Na_2O ratio of 5-11). After the usual treatment phases, sintering
is effected by feeding the articles in a sintering oven, composed of a
refractory material, e.g. concrete, the sintering oven being lined with
a coating of alkaline beta-alumina of the same composition as that of the
articles. Sintering takes place at $1600°-1700°C$ during 30 minutes to 4
hours, followed by cooling the material to ambient temperature.

According to Patent (66/13) of <u>Compagnie Générale d'Electricité</u> ceramic
products are produced from natron-containing beta-alumina powder, which
is dissolved in a polar solvent (methylpropyl ketone and n-pentanol),
the formed suspension being arranged over a support (immersed in the
suspension) and imparting thereto an inverse polarity with regard to
that of the powder, by establishing a potential difference between the
support and the recipient, enclosing the suspension, followed by drying,
compression and sintering.

Patent (85) of <u>Eastman Kodak Company (US)</u> discloses a process for preparing a hot pressed beta-alumina composition by compressing a beta-
alumina powder into a rigid mass. The applied powdery mass is amorphous
and is obtained by a reaction between a solution of an aluminium alco-
holate and an aqueous solution of an alkali metal (bicarbonate, acetate,
hydroxyde, nitrate, carbonate of sodium and/or potassium). The forming
gelatinous mass is then precipitated, the precipitate being heated to
$1200°C$, cooled and crushed into fine powder.

Patent (96/7) of <u>Ford France S.A. (FR)</u> relates to the production of compact ceramic green bodies, which can be sintered into ceramic products
of beta-alumina of high density. The bodies are prepared from a mixture
containing 2.5-4.5 parts of a ceramic composition, 80% of which is aluminium oxide and 5-15% sodium oxide, furthermore 1 part of a binder:
20-35% of polyvinyl pyrolidon with a mol.w. of 20,000-160,000 and 80-65%
of polyethylene glycol. The mixture of the ceramic composition and the
binder is extruded under a pressure of 210-3515 kg/cm^2 at a rate of
6.35-127 cm/min.

Patent (96/8) of <u>Ford France S.A.</u> refers to the sintering of ceramic
green bodies, consisting of polycrystalline di- or multi-metallic oxides,
by enclosing the body in a casing, displaying internal and external wall
portions, the internal wall portion adjacent to the body to be sintered
being made of the same material as this body, while the external wall
portion adjacent to the internal wall portion is composed of a sintered
ceramic material and is practically impermeable and non-reactive to the
said di- or multi-metallic oxides. The body to be sintered and the
said portion of the internal wall are made of beta-alumina or modified
beta-alumina, while the said portion of the external wall is composed
of alpha-alumina. Thereafter the casing and its content are heated to
a temperature and during a time necessary for developing the required
density of the sintered body.

<u>Ford Motor Company (US)</u> disclose in Patent (97/4) a ceramic mass con-
taining polycrystalline beta-alumina of high density and stability with
an electric resistance in view of the conduction of sodium ions lower
than (or equal to) 9 ohm.cm at $300^{o}C$ and a high resistance to rupturing
forces. The ceramic mass is composed of a powder of alumina, a sodium
salt, lithium aluminate ($Li_2O:nAl_2O_3$, wherein n = at least 5). Sintering
of the formed green body takes place at or above $1500^{o}C$, until the
transformation into beta-alumina is completed and nearly the theoretical
density is achieved.

Patent (104/1) and essentially (104/2) of <u>General Electric Corp. (US)</u>
relate to the shaping of beta-aluminium oxide by preparing a suspension
of beta-alumina oxide particles (1-2 microns), the suspension containing
an organic fluid (n-amyl-alcohol) and displaying a dielectricity constant of 12-24 at $25^{o}C$; the beta-aluminium oxide particles are separated
from the suspension by electrophoresis and are sedimented on a charged
electrode (in an electric d.c. voltage field of 100 to 10,000 v/cm).
Next the sediment on the electrode is dried and separated therefrom and
guided through an oxygen-containing atmosphere at a temperature between
1650° and $1825^{o}C$ at a rate of 13-10 cm per minute, thereby obtaining a
sintered beta-aluminium oxide product.

Patent (104/9) of <u>General Electric Company</u> also concerns a process for
producing shaped articles of beta-alumina by preparing a suspension of

beta-alumina particles (with a size between 1 and 5 microns) in an organic fluid that contains 1.0% of magnesium and 0.5% of yttrium oxide. The production process is essentially in accordance with Patents (104/1) and (104/2).

Another Patent (104/21) of General Electric Company concerns a composite article, comprising a first layer of beta-alumina and a second layer of alumina, which is rich in beta"-phase and which adheres tightly to the first layer, that has a thickness corresponding to 1-15% of the total thickness of the article. This article can be formed as a tube, the beta-alumina layer being the inner layer of the tube, while the layer of alumina, rich in beta"-phase forms the external layer of the tube.

Patent (119/3) of K.Hart (DE) relates to shaped elements of polycrystalline alpha-aluminium oxide, containing beta-aluminium oxide, which comprises 0.01-1 w.% of the oxides of vanadium, chrome, manganese, iron, cobalt, nickel, copper, individually or in mixture. The lamellar crystallites of the beta-phase are arranged parallel to one another, with a max. derivation of 15°-20°. The shaped elements may also contain ions of lithium, sodium, potassium, thallium (I) or silver, whereby the ions moving in the ion-conducting planes of the beta-phase can be present individually or within a given group, in any mixture.

Another Patent (119/4) of K.Hart describes the production of sodium-beta-aluminium oxide membranes with statistically oriented microcrystals, the surface of which is coated with a covering layer (50-300 μm) of sodium-beta-aluminium oxide crystals, arranged about perpendicularly to membrane plane. The core of the element displays a higher sodium oxide content than the covering layer, wherein the molar Al-Na-ratio is established between 7 and 11. Both membrane planes are coated with a layer of oriented crystals.

Patent (159/2) of Fried. Krupp GmbH (DE) relates to a shaped ceramic body for use in machining metallic and non-metallic materials and for wear protection, which is free from cracks in the unstressed condition and consists of alpha-alumina; 0.5-35%, preferably 1.20%, by weight of stabilised zirconia, having a degree of stabilisation of 40% to about 100%, preferably 70-90%, and a stabiliser content of 3-9% by weight

calculated on the basis of magnesia and/or CaO. The shaped ceramic body
is produced by mixing the aforementioned amounts of components, which
then are wet-milled to a particle size of less than 3/um, dried at a
temperature up to 150°C and pressed into shapes; the green compacts are

sintered for 0.1 to 3 hours under vacuum or in an atmosphere of protective gas at a temperature of 1450-1750°C and finally cooled to room
temperature under vacuum or in an atmosphere of protective gas.

Patent (163/1) of <u>Kyoto Ceramic Co.Ltd. (JP)</u> concerns a ceramic body for
chromatography, consisting of a calcined molded body of alumina particles,
composed mainly of alpha-alumina, preferably containing a minor amount
(less than 50 w.%) of beta-alumina, and having a narrow pore size distri-
bution range (0.1 to 10 microns). This ceramic body has no substantial
adsorbing property and therefore, when it is used in thin layer chromatography (as a molded plate) or gas chromagraphy (after pulverising),
the tailing phenomenon does not occur and a high separating capacity
can be obtained. The ceramic body is prepared by molding the indicated
composition with a resinous binder and a solvent into the form of a
sheet and calcining it at 850° to 1600°C.

Patent (204/3) of <u>Norton Company</u> provides a body of lower modulus of
elasticity and higher thermal shock resistance than the prior art products
by forming an entirely self-bonded alumina structure, containing coarse
inclusions of silicon carbide particles. Formation of mullite is avoided
by including little or no silica in the mix and by firing at a temperature
low enough that formation of silica by oxidation of the silicon carbide
is avoided. Adequate strength is achieved in such low temperature firing
by including in the mix sufficient very finely divided, active alumina
so that sintering, to provide the required strength in the finished product, is achieved at temperatures of from 1300 to 1535°C. The resulting
product, although having a modulus of elasticity of about 1/3 that of
an equivalent sintered alumina body, suffers no degradation of strength
upon heating as high as 1250°C and has higher thermal conductivity and
thermal shock resistance than prior art bodies containing the same proportions of alumina and silicon carbide. The alumina component must be
a high purity alpha-alumina powder, having an average particle size of
less than 4/um and must be applied in an amount of 10-40%.

Patent (213) of <u>Produits Chimiques Usine Kuhlmann (FR)</u> refers to a process for shaping beta-alumina by adding to a mixture of beta-alumina powder and a binder 4-6 w.% of a powder of silicoboric glass, with a granulometry under 20 microns.

9.3 Other aluminium-based ceramics, aluminates and complexes thereof

According to Patent (11/4) and essentially to (11/5) of <u>Aluminium Pechiney (FR)</u>, aluminium oxide agglomerates can be obtained from an intermediary product, which is derived from the incomplete decomposition of hydrated aluminium sulphate, according to the formula

$$Al_2O_3 \ x \ SO_3, \ YH_2O$$

containing about 3 to 10 w.% sulphur. The intermediary product is compacted under a pressure of 200 kg/cm^2, heat treatment takes place at max. 1500°C. The material is suitable for the production of balls, cylinders, plates, pills, etc.

Patent (36) of <u>Boliden A.B.(SE)</u> concerns a method for recovering usable products from waste products, deriving from the manufacture of aluminium fluoride on the basis of aluminium hydroxide and fluosilicic acid. Silica contaminated with fluorine and aluminium and obtained in the manufacture of aluminium fluoride is dissolved with a strongly basic hydroxide, whereafter the first solution obtained is mixed with a second solution, obtained by dissolving aluminium hydroxide with a strongly basic hydroxide, and with waste mother liquor and optionally also washing water from the manufacture of aluminium fluoride, in such proportions that the pH-value of the mixture is from about 10-14. The silicate content of the waste products supplied is precipitated as a silicoaluminate, which is separated off, preferably by filtration, for further treatment or for direct use, whereafter fluorine, when present in the waste products, is recovered from the filtrate by further precipitation as sodium fluoro-aluminate (Cryolite), and whereafter the filtrate is passed to a recipient or utilised, for example as process water in other processes. The fluorine content can be recovered from the filtrate by adding thereto an aluminium compound in an amount sufficient to precipitate out substantially all the fluorine contained in the filtrate as a fluoroaluminate, which is then separated off.

Patent (81/1) of <u>Dresser Industries, Inc. (US)</u> discloses a refractory
concrete product of increased strength, consisting of 20-30% calcium
aluminate cement, 1-10% silicium carbide of a granulometry, corresponding to 28-35 mash size.

The <u>Henkel K.G. (DE)</u> developed a method according to Patent (121/1) and
(121/2) for preparing amorphous crystallisable sodium-aluminium silicates by the continuous mixing of a sodium aluminate solution with a
sodium silicate solution, the solutions being used in the following
molar proportions:

$$1.5 \text{ to } 9 \text{ } Na_2O : 1 \text{ } Al_2O_3 : 1 \text{ to } 7 \text{ } SiO_2 : 40 \text{ to } 400 \text{ } H_2O.$$

Continuous mixing is carried out in a tubular reactor. The final composition of the solutions mixed together, consisting of amorphous sodium
aluminium silicate, can be crystallised into a zeolitic molecular sieve
of the NaA-type.

According to Patent (201/3) of <u>NL Industries Inc. (US)</u> a novel composition of refractory material is obtained by forming a mixture of 45
to 70 w.% aluminosilicate, 5 to 15 w.% calcined alumina, 15 to 35 w.%
zircon, 1 to 5 w.% clay (bentonite) and 1-5 w.% pyrophyllite, pressing
the mixture in a die and firing the mixture to form the refractory
body. Various alumino-silicate materials can be applied to the composition, such as mullite, sillimanite, synthetic alumina-silica grains,
containing 45-75% Al_2O_3, kyanite, andalusite and the like. The proposed
composition is more resistant to alkaline vapours and dust, forming during
glass melting than the conventional ones.

Patent (224) of <u>Sanac S.p.A. Refrattari Argille e Caolini (IT)</u> provides
a refractory material of siliceous, silicium-aluminium or aluminous
components, the components being mixed for 2-5 minutes,whereafter a
solvent is added to the mixture (in an amount of 3-10% of the refractory), the thus formed paste being agitated for 5-15 minutes and finally
sintered at 1100-1500°C, after having been shaped into bricks. The
mixture also contains additives <u>a</u>) silicates of lithium, potassium,
magnesium in an amount of 0.2-7% of the dry material, or <u>b</u>) sodium
aluminate, sodium phosphate, potassium, lithium, magnesium and phosphoric acid (0.2-7%), while the additive can also be <u>c</u>) a ferrous or

ferric composition, iron, chloride, calcium sulphate, magnesium sulphate, barium sulphate (0.2-7%).

Patent (236) of Société Coordination et Développement de l'Innovation
CORDI S.A. (FR) refers to a mineral polymer compound of the silicoalumi-
nates family, comprised of a solid solution containing a potassium
polysilicate phase having the composition $(y-1)K_2O : (x-2)SiO_2 :$
$(w-1)H_2O$, wherein w is a value which is, at the most, equal to 4,
x is a value comprised between 3.3 and 4.5, y is a value comprised
between 0.9 and 1.6, and a potassium polysialate polymer phase having
the formula:

$$(K)n(-Si-O-Al-O-)n,nH_2O;$$
$$\quad\quad\quad\;\; O \quad\;\; O$$

the potassium polysialate polymer having a diagram to X-rays close to
that of a natural mineral known as kaliophilite $KAlSiO_4$. The process
for producing such polymer and the mineral polymer compound, comprises
the preparation of a reaction mixture based on potassium silicate, on
potassium hydroxide KOH and on aluminosilicate oxide $(Si_2O_5, Al_2O_3)n$,
such that the molar ratios of the reaction products expressed in terms
of oxides, are either comprised of or equal to the following values:
K_2O/SiO_2 : 0.23 to 0.48; SiO_2/Al_2O_3 : 3.3 to 4.5; H_2O/Al_2O_3 : 10 to 25;
K_2O/Al_2O_3 : 0.9 to 1.60, and hardening the reaction mixture at a temperature lower than $120^{\circ}C$. The composition can be used in the manufacture
of molded articles, containing at least a mineral load, as a binder or a
cement, or in the manufacture of articles resisting to thermal shocks,
art objects, moulds and tools.

Patent (246/1) of Sumitomo Electric Industries, Ltd. (JP) reveals a
method of sintering a ceramic material which comprises sintering a
ceramic composition comprising oxides:

$$Al_2O_3, MgO, ZrO_2, Y_2O_3 \text{ or NiO},$$

as a substantial or partial component thereof in an atmosphere of CO
or a mixture of CO and an inert gas which can be argon, helium or
nitrogen. Sintering is carried out at a temperature in the range of
from 1200 to $2000^{\circ}C$.

Patent (271) of UOP Inc., (US) provides a method of preparing pure
alumina extrudate particles of low average bulk density. The alumina
hydrosol acts as both a binder and a lubricant and avoids the need for
extraneous binders and lubricants such as starch, polyvinyl alcohol or
Sterotex, which must be burned from the extruded particles under controlled conditions, particularly where other catalytic components are
present. More particularly, the method consists of (a) forming a mixture
comprising a finely divided alumina and from 2 to 10 w.% alumina hydrosol, based on the weight of alumina plus alumina hydrosol, and adding a
sufficient quantity of an aqueous liquid thereto to produce a dough
extrudable at less than 11 atmospheres; (b) extruding the resulting
dough; and (c) drying and calcining the extrudate. The mixture further comprises a finely divided crystalline aluminosilicate, which is
present in an amount of from 0.5 to 20 w.% of the weight of alumina
plus crystalline aluminosilicate. The mixture may also contain a
platinum group metal. The preferred crystalline aluminium silicate is
mordenite, a zeolite with an SiO_2/Al_2O_3 ratio of from 6:1 to 12:1.

Zirconal Processes Ltd. (GB) reveal in Patent (287/1) a refractory powder
composition of a mixture consisting of two fractions, a first powder
fraction consisting entirely of alumina powder, entirely of mullite powder or entirely of a mixture of alumina powder and mullite powder, and
a second powder fraction consisting of the powdered fusion product of
either: zircon sand and calcined alumina; zirconia, silica and alumina; or zirconia and an aluminosilicate. The first powder fraction
may consist of alumina fines (all finer than 53 microns) and the second
powder fraction is coarser than the first, being the fusion product of
zircon sand and calcined alumina. From this refractory material a
shaped object can be formed by using a binder, which may be a silica
derived from ethyl silicate by amine catalysed hydrolysis and gelation
or by acid catalysed hydrolysis and subsequent gelation. The binder
may be derived from an aluminium halohydrate. The important advantages
of the composition are that it gives a refractory body which has good
abrasion resistance and good resistance to thermal shock. It is especially suited for the refractory components required in the continuous casting of steel.

CHAPTER 10

SILICON-BASED INTERMEDIATES

10.1 Intermediates, mainly comprising silicon carbide

Aktiebolaget Gullhögens Bruk (SE) claim in Patent (6) a synthetic body,
made of at least one hard material and a binder, which penetrates by
infiltration or impregnation, into the pores of the shaped body of hard
material, which can be particulate silicium carbide, corundum or bauxite,
while the binder consists of, for example, calcium silicate or of a glass
of calcium silicate with the following composition: 69 w.% SiO_2; 25 w.%
CaO+MgO; 3 w.% of Al_2O_3; 3 w.% of an alkali or iron oxide.

Patent (16/5) of Annawerk Keramische Betriebe GmbH (DE) describes a
process for producing a polycrystalline shaped body of silicium carbide
with a density of at least 98% of the nominal density of silicium carbide. The shaped body is produced by sintering without the application
of pressure at 1900^0-2200^0C, a mixture of beta-silicium carbide and up
to 3 w.% of boron, without the addition of carbon, in a carbon-containing
atmosphere. The process yields, in spite of the no-pressure sintering,
strength properties of the same order of magnitude as achieved in hot-
pressed silicium carbide bodies.

Annawerk Keramische Betriebe GmbH disclose in another Patent (16/6) a
sintered no-pressure, polycrystalline shaped body with a density of
about 98% of the theoretical one. The body is composed of: at least
92 w.% alpha- and/or beta-silicium carbide, in form of a homogeneous
structure (grain size: max. 10/um); up to 3 w.% boron and about 0.5 -
5.0 w.% of a reducing metal or rare-earth metal, like titanium, zirconium, hafnium, scandium, yttrium, lanthanum and cerium or their mixtures. As SiC powder such powder is used, which displays a specific

surface of 10-20 m^2/g and a granulometry with 95% of the particles under der 1 µm. Sintering takes place at 1900°-2200°C in vacuum or under a protective atmosphere, without pressure.

A ceramic material of dense, sintered silicium carbide is obtained according to Patent (19/4) of <u>Asahi Glass Co. Ltd. (JP)</u> by moulding a mixture of an aluminium source (0.5-35 w.% as expressed in terms of Al_2O_3) and an additive of silicium carbide, the mixture phase being followed by sintering without pressure, in an oxidating atmosphere at a temperature between 1900 and 2300°C.

Patent (22/1) of <u>Bayer A.G.(DE)</u> discloses the production of shaped articles (e.g. pipes, crucibles, blocks, films, coatings, etc.) of a homogeneous mixture of silicon carbide and silicon nitride, obtained by producing a silazane compound by reacting ammonia with at least one carbon-containing halogenosilane at a temperature up to 200°C, converting the silazane compound in the form of a melt or a solution, into a shaped article, and heating the shaped article, in an inert atmosphere, to a temperature between 800°C and 2000°C. The halogenosilane is a compound of the general formula R_nSiX_{4-n}, where R is hydrogen, or an alkyl, alkenyl or aryl group, X is fluorine, chlorine, bromine or iodine, and n is 1, 2 or 3. As shaped article fibres are mentioned, with a diameter of 1-50 µm, which can be used in moulded products as reinforcement or insulating agents.

Patent (32) of <u>Biuro Projektow Przemyslu Metali Niezelaznych "Bipromet" (PL)</u> refers to a method for the anti-corrosive protection of silicon carbide products, preventing the formation of carbon oxide in the product's pores, by enveloping the outer faces of the product with air-tight shields. This shield can be made either of a protective coating deposited on the outer faces of theproduct from the gaseous phase, or of steatite, porcelain, melted ceramic materials or metals. The shield can also be made of substances in melted condition, or as superficial layers of the product impregnated with sealing compounds. Also the carbon oxide concentration in the pores of the product can be decreased by increased number of open pores in the product, enabling oxygen to penetrate to the pores and rarefy the carbon oxide atmosphere. Further

formation of carbon oxide in the pores of the product can be prevented
by formation of a protective atmosphere (N_2) in direct proximity to the
outer faces of the product.

According to Patent (45) of <u>Bulten-Kanthal A.B.(SE)</u> shaped silicium carbide bodies of low porosity can be obtained having a non-porous skeleton
structure of silicium carbide, comprising cavities, which form about
70 vol.% of the body and which are in connection with one another and
are completely filled with a practically non-porous molybdenum silicide
alloy (30-80 vol.% silicium, the rest: molybdenum).

Patent (51/2) of <u>The Carborundum Company</u> describes a sinterable powder
which comprises: a) a particulate ceramic material e.g. SiC, having an
average particle size ranging from 0.10 to 2.00 microns and a surface
area between 5 and 20 m^2/g; b) a carbon source material (e.g. sugar
or a phenol-formaldehyde resin) which will provide between 0.1 and 4.0
w.% of said ceramic material during sintering, and c) a residue from
a solution of H_3BO_3, B_2O_3, or mixtures thereof as sintering aid, substantially uniformly distributed over the particulate portion of said powder
to provide from 0.3 to 5.0 w.% of boron based on the weight of said
ceramic material during sintering.

<u>The Carborundum Company</u> developed several ceramic compositions. Thus,
according to Patent (51/6) sintered ceramic products, containing 55 to
99.5 w.% silicon carbide and 0.5 to 45.0 w.% aluminium nitride, may be
produced by sintering, under substantially pressureless conditions, a
shaped green body, produced by cold pressing a mixture of finely-divided
silicon carbide, aluminium nitride and carbon or carbon source material.
The sintering is conducted in an inert atmosphere at a temperature
between 1900° and 2250°C. Alternatively the products may be produced
by forming a green body from a mixture of finely-divided silicon carbide
and carbon or carbon source material, and sintering the green body under
substantially pressureless conditions in an atmosphere containing from
5×10^{-5} to 1×10^{-3} atmospheres of aluminium nitride at a temperature
between 1900° and 2250°C.

Patent (51/7) of <u>The Carborundum Company</u> refers to ceramic compositions
that may be injection molded and sintered and comprise 65-90% ceramic

powder, 12-30% thermoplastic resin, 2-8% silver wax and 0.1-3% of an
organotitanate which materially reduces the viscosity of the compositions.
The ceramic may be SiC or TiC and B or Be may also be included in the
mixture material. The organo-titanates found useful are represented by
the formula:

$$(R_1 - 0)_m - Ti - (0 - X_z - R_2)_n$$

wherein: a) m is an integer from 1 to 4 and n is an integer from 0 to 4;
b) m+n=4 or 6; c) z is an integer from 0 to 1; d) R_1 is aliphatic
containing from 1 to 8 carbon atoms; e) X is independently selected
from the group of phosphite, phosphate and pyrophosphate; f) R_2 is
aliphatic containing from 8 to 25 carbon atoms.

The Carborundum Company presents in Patent (51/11) a compact body of
silicium carbide, which is highly resistant to temperature changes.
This body consists of a hot-pressed mixture of silicium carbide with
0.2-2.0 w.% aluminium diboride and carbon. The body is obtained by
hot pressing under a high pressure for a period of time during which
the body obtains 99% of the theoretical density of massive silicium
carbide.

A similar compact body is revealed in Patent (51/12) of The Carborundum
Company which consists of a pressure-free sintered mixture of silicium
carbide, furthermore 0.3-3 w.% of an additive of boron nitride, boron-
phosphide, aluminiumdiboride or their mixture, and also carbon in an
amount of 150-500 w.% and in form of an organic compound (added before
sintering).

A sinterable powder, described in Patent (51/13) of The Carborundum
Company comprises: a) a particulate ceramic material e.g. SiC,
having an average particle size ranging from 0.10 to 2.00 microns and
a surface area between 5 and 20 m^2/g; b) a carbon source material
(e.g. sugar or a phenol-formaldehyde resin) which will provide between
1.0 and 4.0 w.% of the said ceramic material during sintering, and
c) a residue from a solution of H_3BO_3, B_2O_3, or mixtures thereof as
sintering aid, substantially uniformly distributed over the particulate
portion of said powder to provide from 0.3 to 5.0 w.% of boron based on
the weight of said ceramic material during sintering.

Patent (51/18) of the same company relates to a ceramic composition,
suitable for injection moulding, which contains α-silicium carbide,
as the major component of the composition, further 0.2-5.0 w.% of a
sintering-promoting agent of boron or beryllium (elementary beryllium,
beryllium compounds, elementary boron, boron compounds). This agent
should preferably contain boron carbide, 14-30 w.% of a thermoplastic
resin (polystyrene) and another ingredient in form of oil or wax, the
volatilisation temperature of which is below that of the resin.

A silicium carbide powder, claimed in Carborundum Patent 51/22 contains
beryllium or a beryllium-containing compound, which acts as densifying
agent for a ceramic material. The sintering of the powder takes place
in a beryllium-containing atmosphere, wherein the partial pressure of
beryllium corresponds to (or is higher than) the equilibrium vapour
tension of beryllium in the powder. The silicium carbide powder contains about 0.03-1.5 w.% beryllium and in the sintering atmosphere
there is also an inert gas present.

A process for producing silicium carbide of high density is described
in Patent (51/23) of The Carborundum Company according to which a mixture
of silicium carbide (containing about 6.0 w.% of elementary carbon or a
carbon source) is formed into a blank and sintered (essentially without
pressure) in a boron-containing atmosphere, thereby obtaining a product
of a density of 85% of the nominal value. Boron is applied in the form
of boron carbide and is fed in the sintering chamber in form of a paste
prior to sintering.

Patent (51/25) of The Carborundum Company reveals a process for sintering silicium carbide powders. In order to obtain a ceramic material
of high density, sintering is carried out in a boron-containing atmosphere, in which an inert gas (nitrogen, argon, helium) is also present.
The silicon powders have a boron content of 0.1 to 5.0 w.%. The partial
boron pressure in the sintering atmosphere amounts to at least 10^{-2}Pa.

Patent (56/3) of Ceraver S.A. (FR) reveals a process for producing
silicium carbide elements by sintering without pressure. A carbide
powder is exposed to thermal treatment under vacuum, which then is mixed
with a nitride powder, with less than 0.15% boron content, whereafter

a carbonisable material is added to the mixture, producing 0.2-4 w.%
carbon and a casting is produced from the mixture, which then is sintered at 1900-2250°C. The said nitride is the product of nitriding a
powder mixture of silicium and aluminium.

In the process, described in Patent (56/4) of Ceraver S.A. a ceramic
article is obtained of sintered silicium carbide by forming a mixture
of a carbide powder, or a sintering promoting agent on the basis of
boron, beryllium or aluminium, of a thermoplastic resin or wax, which
promote injection moulding of the mixture, furthermore of an organic,
carbonisable material. Polystyrene can be used both as a carbonisable
material and as promoting agent. The mixture is sintered between 230°
and 330°C and kept in this temperature range for a time necessary for
removing the wax therefrom and for reticulating and oxidising polystyrene.

Patent (66/11) of Compagnie Générale d'Électricité concerns a ceramic
material of silicium carbide and an additive, selected from SiB_4, SiB_6,
AlB_2, AlB_{12}, BN or Al_4C_3 and applied in an amount of 0.5-5% of the
silicium carbide weight. The silicium carbide ceramic material displays
a 95% density.

Patent (66/14) of Compagnie Générale d'Électricité refers to a process
for producing a ceramic product of dense silicium carbide by preparing
a mixture of alpha-silicium carbide powder, boron powder and a carbonaceous substance, the mixture being sintered in neutral atmosphere at
about 2100°C. The preparation of the mixture includes a thermal treatment under vacuum at 1200-1400°C in such a way that oxygen and silicium
oxide can be separated from the powder. Boron is introduced into the
mixture prior to heat treatment. The carbonaceous component of the
mixture is a phenol lacquer, which is dissolved in water and added to
a powder after heat treatment. Before sintering, the mixture is
crushed in a ball mill, the balls therein being coated with polytetrafluorethylene.

According to Patent (71/4) of Daimler-Benz A.G. (DE) a honeycomb structure
can be produced of silicium nitride, containing a plurality of thin-
walled channels. This structure is obtained by arranging a layer of

silicium powder (by thermal spraying) on a graphite base; graphite
strips of trapezoid cross-section are then placed with a regular spacing
in the silicium layer (with their base downward), whereafter a silicium
layer will be spread (by thermal spraying) over the graphite strips and
the non-covered zones of the first silicium powder layer.

Patent (90/2) of Elektroschmelzwerke Kempten GmbH (DE) provides compact
polycrystalline shaped bodies of alpha-silicium-carbide with small
amounts of aluminium carbon, nitrogen and a very small amount of residual oxygen, whereby only carbon can be identified as a separate phase.
The shaped bodies display high thermal stability and bending strength of
at least 500 N/mm^2 up to 1600°C and a transcrystalline fracture mode.
The shaped bodies are obtained by cold pressing submicron powder particles of alpha-silicium carbide, aluminium powder, or a non-oxide aluminium composition, a carbon-containing additive like soot, phenol-formaldehyde condensation products, coal tar pitch, into a shaped form which
then is sintered without applying pressure at 2000-2300°C.

Patent (90/3) of Elektroschmelzwerke Kempten GmbH reveals the production
of compact polycrystalline alpha-silicium carbide, with the addition of
small amounts of aluminium, nitrogen, phosphor in a homogeneous, mono-
phase micro-structure,while the additives are present in the form of a
solid solution in the crystal lattice. The mixture of alpha-SiC,
aluminium (powder), aluminium nitride and/or aluminium phosphide is hot-
pressed at 1850°-2300°C, under a pressure of 100 bars (10 MPa).

Patent (90/4) of Elektroschmelzwerke Kempten GmbH refers to compact,
shaped bodies of polycrystalline beta-silicium carbide with the same
additives and under the same conditions as applied in Patent (90/3).
The mixture is hot-pressed between 1750°C and 2200°C, under a pressure
of at least 100 bars (10 MPa).

Elektroschmelzwerke Kempten GmbH also present in Patent (90/5) a practically pore-free shaped body of polycrystalline alpha- and/or beta-
silicium carbide in form of a single-phase homogeneous micro-structure,
with grains of 8 /um, of pure SiC powder and not more than 0.1 w.% of
metallic impurities. The products are obtained by isostatic hot pressing
in a vacuum-proof casing at 1900°-2300°C under a pressure of 100-400 MP

(1-4 kbar) in high-pressure autoclaves in an atmosphere of inert gases. The SiC powder, fed into the cavity, is compacted by vibrations.

General Electric Company reveal in Patent (104/5) a process for producing a dense ceramic body of silicium carbide from a dispersion of a powder (of submicron-size particles), containing silicium carbide, boron carbide, a carbonaceous additive (with an equivalent quantity of 0.1 - 1.0 w.% carbon), the pulverulent mixture then being formed into a green body and sintered in an inert atmosphere (argon, helium, nitrogen, hydrogen) at 1900°-2100°C.

In Patent (104/11) General Electric Company claim a process for the production of silicium carbide products by forming a mixture (of submicron particles), containing: silicium carbide, composed of beta and alpha phase, the latter being in a 0.05 to 5 w.% relationship with regard to the beta-phase; boron carbide (0.3-3.0 w.% of the total amount of silicium carbide), further: a carbonaceous additive (free carbon, organic carbonaceous substance). From this mixture a green body is formed, which then is sintered at 1950-2300°C, in inert atmosphere, at atmospheric or subatmospheric pressure, to obtain a product of a density of 80% of the nominal value.

Patent (104/14) of General Electric Company concerns a machinable casting with a volumetric mass of 1.6-2.7 g/cm^{3}, obtained by infiltration and reaction of a silicium smelt and a mixture, which contains 45-90% particulate carbon and silicium carbide in an equal amount; 10-50% of a particulate inorganic material, that is inert with regard to molten silicium and that consists essentially of boron nitride. The casting contains the carbon component in the form of carbon fibres, or as a mixture of carbon fibres and silicium carbide particles.

According to Patent (104/15) of General Electric Company a polycrystalline sintered body can be obtained from silicium carbide and boron carbide, by preparing a dispersion, containing sub-micron size particles of beta-silicium carbide, boron carbide (10-30 w.% of the total amount of silicium carbide and boron carbide), furthermore a carbonaceous additive, containing either free carbon or a carbonaceous organic substance, which is completely decomposed between 50° and 1000°C in order to release free carbon and a gaseous decomposition product; the dispersion is formed into a green body, which is sintered at a temperature between 2000°C and a degree lower than the fusion point of the eutectic system of silicium carbide and boron carbide. The final product has a density of 85% of the nominal value.

In Patent (115) Groupement pour les Activités Atomiques et Avancées (FR) disclose a refractory ceramic material, obtained by burning at a temperature lower than the sintering temperature ; the material consisting of cement and a mixture of aggregates, which contains at least a ceramic fibre mixed in a proportion of more than 10% with regard to the other refractory aggregates, which may contain tabular aluminium oxide, corundum, stabilised zirconium, lime zirconate or silicium carbide.

Fa. Hermann C. Starck (DE) present in Patent (122) a process for the production of sintered, refractory, compact, shaped bodies of silicium carbide with a fine crystalline structure, comprising no coarse-crystalline 6H-SiC-plates. The shaped bodies are obtained by preparing a homogeneous dispersion of fine powder, that contains 90,0-99,6 w.% of beta- and alpha-phase silicium carbide (or mixtures thereof), to which there is added an aluminium-containing additive in an amount of 0-4 w.% (Al), a carbonaceous additive, in an amount of 0.4-6 w.% (carbon), the powder mixture being compacted by compression, casting, or any other suitable method. Sintering takes place in inert or reducing and Al-containing atmosphere at 1850-2100°C, without pressure and under conditions ensuring a rel. density of 85% of the theoretical density of the sintered body.

Patent (123/2) of Hiroshige Suzuki (JP) relates to the production of silicon carbide powder having a sub-micron grain size, which can form a sintered body having a density of more than 93% of the theoretical density and containing more than 90% of beta-silicon carbide. The process for producing such powder comprises: reacting 1.7 to 2.1 w.% of silicon monoxide with 1.0 w.% of finely divided carbon at a temperature of 1200 to 1500°C under a reduced pressure of lower than 10 mmHg. The reaction is carried out under an inert gas. The finely divided carbon applied in the process is carbon black, produced by pyrolysis or carbonisation of a hydrocarbon vapour or an organic compound.

Hitachi Ltd. (JP) developed in Patent (124/5) an electrically insulating
substrate comprising a sintered body, which consists of 0.1 to 3.5 w.%
of beryllium and silicon carbide as the principal components, while
silicon carbide contains up to 0.1 w.% of aluminium, up to 0.1 w.% of
boron and up to 0.4 w.% of free carbon. The body is so sintered that
a density of at least 90% relative density with respect to the theoretical density of said silicon carbide is obtained and has a thermal
conductivity of at least 1.67 J/cm.sec.^{o}C at 25°C and a coefficient of
thermal expansion of up to 4 x 10^{-6}/^{o}C from 25°C to 300°C. From the
aforementioned components an electrically insulating substrate is made,
by mixing 0.5 to 14 w.% of beryllium oxide powder with silicon carbide
powder containing up to 0.1 w.% of aluminium, up to 0.1% of boron and
up to 0.4% of free carbon; followed by pressure molding and sintering
at 1850-2500°C in a non-oxidising atmosphere.

According to Patent (124/6) of Nakamura Kosuke (JP) a mixture is prepared
for producing an electrically insulating silicon carbide sintered body,
comprising silicon carbide containing 0.4 w.% or less of free carbon and
0.1 w.% or less of aluminium as impurities, as a major component and at
least one beryllium compound which can be converted to beryllium, beryllium oxide, beryllium carbide or beryllium nitride by decomposition with
heating. The amount of beryllium compound is 0.05 to 7.5 w.% in terms
of the amount of beryllium based on the amount of silicon carbide, the
beryllium compound being at least one member selected from the group consisting of $Be(OH)_2$, $Be(NO_3)_2$, $BeSO_4$, $BeSO_4 \cdot 2NH_3$, $BeCO_3$, $Be(HCO_2)_2$,
$Be_4O(HCO_2)_6$, BeC_2H_4, $Be_4O(CH_3CO_2)_6$, $(NH_4)_2O \cdot BeO \cdot 2C_2O_3$, $(NH_4)_2Be(SO_4)_2$,
and $Be(CH_3COCHCOCH_3)_2$.

Patent (126) of Ibigawa Electric Industry Co.,Ltd. (JP) provides a
method for producing a silicon carbide sintered body,having a density
of at least 2.4 g/cm^2 by preparing a mixture, consisting mainly of
silicon carbide with an average particle size of not more than 3.0
microns, and a temporary binder; shaping the mixture into a green body
and sintering the green body at a temperature of 1750°C-2100°C.
Before sintering, silicon carbide is contacted with either hydrofluoric
acid or anhydrous hydrofluoric acid and the treated silicon carbide is
held in a non-oxidising atmosphere after the contact treatment up to
the completion of sintering. The treatment with hydrofluoric acid may

be before, or after mixing or after pressing. The mixture may also contain up to 5% sintering acids such as Al, W, Mg, Fe, B or B_4C.

International Ceramic Ltd. (GB) disclose in Patent (140) a method of forming a profiled body of a ceramic material, having high strength and high temperature characteristics, the method comprising the steps of mixing silicon with a predetermined quantity of aluminium nitride in powdered form, subjecting the mix to injection moulding so as to form the profiled body, converting the silicon to silicon nitride, introducing a material to voids created in the silicon nitride, and firing the profiled body, whereby reaction of the profiled body and the introduced material and silica which is inherent at boundaries in the formed silicon nitride forms a body having the high strength characteristics free of unconverted silica to prevent strength loss at high temperature. In a modification of this method, the profiled body may be internally oxidised so as to fill the voids with silica from the silicon nitride. The body may, for example, be infiltrated with a proprietary alumina sol, being then dried and fired to at least $1800^{o}C$ in a controlled atmosphere. Alternatively, the silicon nitride is infiltrated with an aqueous solution of aluminium sulphate. Then before or after drying, the body is treated in ammonia solution so that the aluminium hydroxide is precipitated in the pores of the body. Subsequent heat treatment decomposes the hydroxide to aluminium oxide. The body is then fired to at least $1800^{o}C$ in a controlled atmosphere to form a sialon.

The object of Patent (142) of Ishizuka Garasu K.K. (JP) is a composite material, essentially consisting of silicium carbide particles, bonded together by a glass ceramic. The size of silicium carbide and glass-ceramic particles is between 5 and 150 microns, while the volume ratio of these particles is established between 80/20 and 40/60 (in the above order). The glass ceramic is composed of $Li_2O-Al_2O_3-SiO_2$; $MgO-Al_2O_3-SiO_2$ or $Li_2O-MgO-Al_2O_3-SiO_2$.

Japan Metals & Chemicals Co.,Ltd. (JP) claim in Patent (143) shaped products of sintered silicium carbide of high resistance, which are composed of 0.027-11.300% of atoms of at least one rare-earth oxide; 0.006-11.500% of atoms of aluminium oxide and boron oxide(s), while the rest of the composition essentially consists of silicium carbide.

Patent (149/3) of <u>Kennecott Corp. (US)</u> relates to the sintering of green
bodies of refractory materials, the green bodies being placed in an oven,
through which oxygen-free gas currents are passed. A feature of the
described process is that preheated gas is fed into the oven for the
direct transfer of heat into the shaped body, the preheating temperature
being higher than 4000°C, thereby heating the body to a temperature
between 1900° and 2200°C. The green body consists of alpha-silicium
carbide and the preheated gas for direct heat transfer contains nitrogen
or argon.

Patent (149/4) of <u>Kennecott Corp.</u>provides dense, strong, composite
materials, produced from mixtures of silicon and silicon carbide. The
composite materials consist of a mixture of finely-divided silicon
carbide in a substantially continuous matrix of silicon, and contain
from about 40 to 60 vol.% silicon carbide and from about 60 to 40 vol.%
silicon. The composite materials are obtained by initially preparing
a finely-divided silicon carbide starting component and forming the
component into a green body of the desired shape. The green body is
subsequently impregnated with molten silicon in the absence of any substantial amount of carbon. Suitably, a thermoplastic resin binder may
be added to the silicon carbide starting material to aid in forming
the green body. The impregnation step may be carried out by surrounding the green body with finely-divided silicon metal and heating the
covered body to a temperature above the melting point of silicon. The
composite materials described have flexural strength of over about
90,000 psi at 1100°C, elastic modulus of about 38 million psi at room
temperature; they are essentially non-porous and have densities ranging
from about 2.6 to about 2.8 gcc.

According to Patent (155) of <u>Koppers Fabriken Feuerfester Erzeugnisse</u>
<u>GmbH (DE)</u> refractory bricks are obtained from a mixture containing
30-90 w.% silicium carbide and 10-70 w.% corundum. The fine fraction
of the mixture consists only of corundum. The mixture may also contain up to 5 w.% clay as a binder. The bricks can be used in building
and lining coke ovens.

Patent (162/4) of <u>Kurosaki Refractories Co.,Ltd. (JP)</u> concerns a flaky
beta-silicium carbide consisting of high-grade beta-silicium carbide,

obtained from an organic silicium polymer, which contains carbon and silicium atoms as the main components. The flaky beta-silicium carbide can be used in producing ceramic products of laminated structure, furthermore refractory materials, displaying a high resistance to heat shocks, thermal aging and oxidation.

Kyoto Ceramic K.K. (JP) present in Patent (163/3) a process for producing silicium carbide bodies from polycarbon silane, by polymerising organosilicium compositions into polycarbon silane, which cannot be dissolved or fused, since its smelt-temperature is higher than that of thermal dissociation. Next the polycarbonsilane is powdered, the powder being transformed into silicium carbide powder at 600°-2200°C in a non-oxidising atmosphere, and then being shaped and sintered. The obtained silicium carbide is essentially in amorphous state.

Morgan Thermic S.A.(FR) developed in Patent (182) a refractory cement composition, containing hydraulic cement and a liquid binder on the basis of aluminium phosphate, which is obtained by mixing it with aluminium oxide and sirup-like phosphoric acid. The binder is applied in an amount of 15-35 litres per 100 kg solid mixture, that contains 10-50 w.% hydraulic cement; 0-6 w.% plasticising clay; 30-80 w.% silicium carbide; 10-30 w.% ferro-silicium or silicium (95%). From this material plates and ceramic refractory heat-conducting elements are produced.

Patent (183) of H.R.Moskopf (DE) describes a process for producing refractory, stamping, injection and casting masses for lining smelting equipment, cyclones, refuse incinerators and the like, from a starting mixture of 17-95 w.% SiC particles of 0-4 mm size; 1-20 w.% SiN particles of 0-01 mm size and 1-15 w.% tempered, condensed aluminium phosphate in powder form of particles of 0-1 mm size, further 2-8 w.% refractory bonding clay and 0.5-1.5 w.% of an organic binder, for example cell pitch powder.

NGK Insulators Ltd. disclose in Patent (190/2) a process for gluing together at least two ceramic silicium elements by inserting a first silicium ceramic element into a second one for obtaining the required arrangement, the first element displaying a firing shrinkage, which

is less considerable than that of the second element, both elements
being so sintered that they will be tightly bonded into an integrated
body.

Patents (194/1), (194/2) and (194/6) of Nippon Crucible Co.Ltd.(JP)
refer to an active silicon carbide powder, which contains a boron component as boron carbide or as a solid solution thereof in a uniformly dispersed state in an amount of about 0.2 to 10 w.%, calculated as boron
carbide. To produce the active silicon carbide powder, carbon powder
of a particle size of 20/um or less, metallic silicon powder and a boron
oxide powder are mixed in specified percentages and the resulting mixture is heated in an oxidising atmosphere, containing about 0.3 to 35
volume % of oxygen to induce, at 800-1450°C, a reaction between the
components of the mixture, the reaction being completed substantially
instantaneously.

The object of Patent (195/2) of Nippon Gaishi K.K. (JP) is a silicon
carbide powder having a high sinterability and which can be produced
by thermally decomposing an organosilicon polymer containing silicon,
boron and oxygen as main skeleton components and having groups containing carbon in the side chain at a temperature of 1300-2000°C under
vacuum, under a reducing atmosphere or under an inert gas atmosphere,
heating the thermally decomposed powdery product at a temperature of
500-800°C under an oxidising atmosphere, and then treating the powder
with acid, containing at least hydrofluoric acid. The silicon carbide
powder can be formed into a sintered body having a high density and
bending strength by sintering the powder together with carbon or a
mixture of carbon and boron under vacuum, under a carbon monoxide gas
atmosphere or under an inert gas atmosphere.

Patent (198) of Nippon Oil Seal Industry Co.Ltd. (JP) relates to the
production of shaped silicon carbide articles. The proposed process
uses a carbon starting material, which has a true specific gravity
(i.e. the specific gravity of the carbon material per se) of not more
than 2.1, from which articles of the desired form are shaped by any
suitable conventional technique, to provide a shaped carbon article,
having a porosity of from 20% to 40%. The shaped carbon articles are
then treated with silicon monoxide gas at a temperature of from 1800°

to 2100°C to convert at least the surface of the articles into silicon carbide. The carbon material applied has a coefficient of thermal expansion of from 3 to 6 x 10^{-6}/°C.

Patent (199/1) of <u>Nippon Steel Corporation and Harima Refractory Co.Ltd. (JP)</u> concerns a refractory mixture of non-sintered silicium carbide, which is composed of about 95-99.5% silicium carbide containing 1.5-8% particles under 1 μ and 0.5-5% ultrafine silicium oxide powder, containing at least 50% particles under 1 μ after coagulation. The mixture also comprises 0.5-4% metallic silicium and 0.3-4% powder of a thermoplastic resin (phenol or novolac) resin for thermal hardening.

According to Patent (204/14) of <u>Norton Company</u>, articles of dense silicium carbide can be obtained in complex forms by preparing a green body of powdery silicium carbide (with particle sizes of less than 3 microns and between 30 and 170 microns), the green body essentially consisting of silicium carbide (85-99 w.%) and a metal selected from boron, aluminium, iron (1-15 w.%), the green body first being sintered and partly densified at a lower temperature, while after shaping of the green body, another heat treatment is effected at 1850-2150°C.

Patent (208) of <u>Pickford Holland Co.Ltd (GB)</u> provides improvements in or referring to bonds for refractory materials, more particularly the method for achieving this aim comprises: admixing a first, suitably graded component of refractory material, a second silicon powder component and a third powdered graphite component, and firing this mixture in an atmosphere of nitrogen, whereby the first refractory component is bonded to the third graphite component by the product of the reaction between silicon and nitrogen. Firing takes place between 1100° and 1350°C.

Patent (217/1) of <u>Research Institute for Iron, Steel and other Metals of the Tohoku University (JP)</u> describes the production of fibre composites, composed of continuous silicium carbide fibres and metallic silicium. The cavities in a staple of essentially continuous silicium carbide fibres are filled with molten metallic silicium, realising a close bond between the fibres and metallic silicium, which is present in an amount of 5 to 35 w.%.

In Patent (217/5) Research Institute for Iron, Steel and other Metals of the Tohoku University claim the production of composite refractory materials, which are reinforced with silicium carbide fibres. The material is obtained by forming a vein of powdery ceramic material (with a particle size under 100/u) and of silicium carbide fibres, obtained by firing spun fibres of organosilicium of high mol.weight, the composition thus obtained being compressed and sintered.

Another Patent (217/6) of the aforementioned Research Institute and also Patent (251) granted to the Tohoku University refers to a method for obtaining sintered profiled articles of silicium carbide by mixing the initial material with at least one binder of the group containing compositions a) with only Si-C bonds; b) Si-H + SiC bonds; c) Si-Hal bonds; Si-Si-bonds and organosilicon compositions of high molecular weight, wherein silicium and carbon are the main structural elements. Forming of the articles takes place at 600°-1700°C in oxidising atmosphere to remove free carbon. The binder is applied in an amount of 0.3-30 w.%.

Rosenthal A.G.(DE) disclose in Patent (221/1) a process for producing silicium-containing foam bodies by treating silicium powder (particle size: under 63/um) with diluted caustic soda, the obtained stable foam then being poured into forms and - through firing - in nitrogen-containing atmosphere, converted into silicium nitride, or through firing in carbonising atmosphere into silicium carbide. The addition of caustic soda of 0.3-5% concentration ensures the production of a tough silicium dross. The cell structure of foam can be modified by the varying amount of caustic soda, water, particle size of silicium powder, reaction time, etc. The silicium dross may contain 0.2-1% ammonium alginate or another distillable temporary binder.

Patent (229/1) of Shinagawa Refractories Co.Ltd.(JP) refers to refractory aluminium oxide - silicium carbide compositions, consisting essentially of an intergrown structure of corundum (50-98 w.%) and silicium carbide (2-35 w.%). Such refractory materials can be obtained by adding fine aluminium powder to the silicium dioxide - aluminium oxide material, followed by blending, shaping and drying, the green body thus obtained being fired in a carbonoxide gas atmosphere. The SiO-AlO-material is produced by fusion or from the gas phase, in form of silicium dioxide, roseki-agalmotolite, bentonite, flint clay, mullite, etc.

Patent (229/2) of <u>Shinagawa Refractories Co.Ltd.</u> provides refractory
lining material for vats for iron smelts, obtained by mixing expandable
silicium-containing refractories with 5-50 w.% silicium carbide, 2-30 w.%
silicium, aluminium and chrome in powder form. The silicium-containing
refractories include: pyrophyllite, quartzite, silicon sand, which may
contain 65-98 w.% SiO_2.

The object of Patent (232/3) of <u>Sigri Elektrographit GmbH (DE)</u> is to
produce refractory articles, consisting of graphite, carbon, silicium
carbide for the lining of metallurgical ovens, casting forms and chutes.
The components are applied in following amounts: 5-60 w.% graphite;
10-65 w.% carbon; 10-35 w.% silicium carbide and 10-30 w.% silicium
nitride. Silicium nitride displays a grain size of $<$ 0.1 mm.

In Patent (234) of <u>Smiths Industries Ltd. (GB)</u> a semi-conductor body is
described, destined for example for use as igniter with superficial
discharge. The body is obtained by mixing together silicium carbide,
a silicate or a substance capable of forming a silicate, for example
aluminium silicate (glass). Before compression and sintering the mixture is heated to 1150-1300°C in order to permit the coating of silicium carbide particles with silicium oxide.

Patent (238/1) of <u>Société Européenne de Propulsion S.A. (FR)</u> refers to
the production of a porous, densified, carbonised body of a material,
composed of silicium carbide, boron carbide, transition elements, nitrides of silicium and boron, silicides of metals, diborides of transition elements. The porous body is densified by carbon infiltration.

<u>UBE Industries Ltd. (JP)</u> developed, according to Patent (263/5) a process
for producing a sintered ceramic body, which comprises: heating a semi-
inorganic block copolymer at a temperature of from 500 to 2300°C in an
environment of vacuum or inert gases, reducing gases or hydrocarbon gases,
this copolymer comprising polycarbosilane blocks, having a main-chain
skeleton composed mainly of carbosilane units of the formula $\{Si-CH_2\}$
and titanoxane units of the formula $\{Ti-O\}$; and shaping the heated product, and simultaneously with, or after, the shaping of the heated product, sintering the shaped product at a temperature of from 800°C to
2300°C in an environment of vacuum or inert gases, reducing gases or

hydrocarbon gases; and a sintered ceramic body consisting substantially
of Si, Ti and C and optionally of O, said sintered body being composed
substantially of (1) an amorphous material consisting substantially of
Si, Ti and C and optionally of O, or (2) an aggregate consisting substantially of ultrafine crystalline particles of β -SiC, TiC, a solid
solution of β -SiC and TiC and TiC$_{1-x}$, wherein $0 < x < 1$ and having a
particle diameter of not more than 500 $\AA$, or (3) a mixture of said
amorphous material (1) and said aggregate (2) of ultrafine crystalline
particles.

Another Patent (263/7) of UBE Industries refers to a process, similar
to that described in (263/5) using however for the composition of a
sintered body Si, C, and instead of Ti, Zr.

The United Kingdom Atomic Energy Authority (UKAEA),   developed a
series of refractory ceramic materials, of high strength, low thermal
expansion coefficient. Thus, in Patent (267/1) there is disclosed a
method for the production of an electrically conducting material, comprising a matrix of silicon nitride having dispersed therein an electrically conducting phase of silicon carbide. The method comprises: subjecting to nitriding conditions a mixture,comprising silicon and carbon, having a particle size of not greater than 50 microns and capable
of reaction with silicon under the said conditions to give the electrically conducting phase, so that a part of the silicon is nitrided to
form the silicon nitride matrix and the other part of the silicon reacts
with the component to give the electrically conducting phase, in which
method the conditions and the proportion of the silicon and the component in the mixture are such that free component is absent in the
material produced. The mixture is in the form of an artefact,before it
is subjected to the nitriding conditions. Carbon is in the form of
colloidal graphite.

Another Patent (267/19) of UKAEA refers to a method of modifying mechanical properties (hardness, indentation, fracture) of a surface of a
silicon carbide body by implanting into the surface ions of nitrogen.
The ion dose and energy are applied to such an extent that a material
layer of increased plasticity can be formed. For example, the ion dose

can be greater than 5×10^{17} ions/cm^2, and implantation can be effected
with an energy of between 5 and 500 KeV.

The object of Patent (283) of P.Wecht (DE) is a process for producing
SiC-bonded SiC-shaped bodies by forming the SiC binders in the shaped
body proper, whereby the SiC particles are mixed with crushed rice
husks or substances consisting of Si and C in equal amounts, prior to
shaping. The addition of rice husks (in an amount of up to 50 w.%)
takes place at 300°-900°C.

10.2   Intermediates based on silicon nitride

Patent (1/1) of Advanced Materials Engineering Ltd. (GB) relates to a
method for the production of foamed refractory material, by mixing together precursors, which are liquid or capable of conversion to the
liquid state at moderate temperatures and which react to give a foamable resin; a blowing agent for the resin and silicon to form a green
foam; debonding the green foam and heating it in an atmosphere of or
containing nitrogen in order to convert the silicon to silicium nitride.
The precursors applied are such as to produce a polyurethane resin or
a phenolic resin. A further ingredient is incorporated in the mixture,
which is capable of conversion to a refractory material, e.g. carbon,
which is converted to silicon carbide.

Another Patent (1/3) of Advanced Materials Engineering Ltd. reveals a
method of treating a permeable refractory material or metal comprising:
diffusing at least two fluids into the permeable material in such a
manner that neither fluid meets any other except within the material
and causing a chemical reaction between at least two of the fluids within
the material to produce a deposit of refractory material with the permeable material, which can be in the form of a shaped, partly shaped
or unshaped body. The permeable material is silicon nitride, silicon
carbide, aluminium oxide or another material. The shaped body can be
used as a heat-exchanger matrix.

Patent (4) the Agency of Industrial Science and Technology (JP)
describes a process, wherein silicic acid (anhydrous form), SiO$_2$,

aluminium Al and nitrogen N are reacted together according to the following formulas, whereby a sintered product composed predominantly of a solid solution of β-Si_3N_4. α-Al_2O_3 or β-Si_3N_4.AlN series is produced:

$$3SiO_2 + 4Al + 2N_2 \rightarrow Si_3N_4 + 2Al_2O_3$$

or

$$3SiO_2 + 6Al + 3N_2 \rightarrow Si_3N_4 + 2Al_2O_3 + 2AlN$$

The process comprises: heating a powdery mixture of 80 to 40 w.% of non-crystalline or crystalline silicon dioxide and 60 to 20 w.% of aluminium in a nitrogen atmosphere, the heating being initiated at $1200^{o}C$ and being terminated at a temperature in the range of from $1600^{o}C$ to $2000^{o}C$, to form a sintered product.

Patent (9/1) of Allmanna Svenska Elektriska A.B. (SE) concerns the production of silicium nitride powder or of a shaped body prepared from such powder under compression at $1600^{o}C$, by isostatic pressure of at least 100 MPa, with a gas as pressure agent and by including the powder (body) into a casing, that is gas-impermeable. The casing, containing the powder (body) is exposed to a vacuum treatment, prior to introducing the powder (body) therein. The casing is made of glass, displaying a high softening point.

Allmanna Svenska Elektriske A.B. present in another Patent (9/2) a process for producing isostatically compressed masses of silicium nitride powder, enclosed in a casing under such temperature and pressure conditions, which ensure that the powder particles are welded together. The process includes the formation of a blank of the powder, placing it in a glass casing of a first composition and coating it in a glass powder of a second composition of higher softening point. Then the casing is evacuated and introduced in an oven, which is heated to 1700°-$1800^{o}C$ for forming the final product.

Annawerk Keramische Betriebe GmbH (DE) disclose in Patent (16/1) a process for improving the bending strength (mainly at high temperature) of silicium nitride bodies, by nitriding a porous silicium powder Bank. More in particular the shaped silicium nitride body will be

superficially impregnated with a solution (smelt) containing magnesium oxide or magnesium silicate or substances, which develop them at
high temperatures, the impregnated bodies then being hot-pressed at
1600°-1800°C under a pressure up to 500 kp/cm^2. The impregnating
agent (a salt solution) is applied by brushes to the shaped body.

Patent (20/1) of ASEA A.B. (SE) refers to the production of silicium
nitride articles under isostatic pressure from a pre-shaped powder of
silicium nitride with the aid of a pressure fluid at a temperature
required for the sintering of silicium nitride. Prior to pressing,
the pre-shaped mass is degazified. On the pre-shaped powder mass an
internal porous layer of a first material (glass) is placed, and on
this porous internal layer, in turn, an external layer of a second
material is arranged, which consists of a low fusion point glass.
The porous internal layer can be transformed, with cooperation of the
second material, into a layer which is impermeable to the pressure
fluid, by heating at a temperature, which is higher than the sintering
temperature of silicium nitride, while the porous external layer can
be transformed into a layer, which is impermeable to the pressure liquid, by heating to a temperature, which is lower than the temperature
required for the porous internal layer. The first material is composed
of high fusion point glass or of high fusion point metal.

A process for producing silicium nitride articles according to Patent
(20/2) of ASEA A.B. consists of compressing a powder (or pre-formed
body) of silicium nitride at 1600°C. Prior to compression the silicium
nitride powder is mixed with yttrium oxide (or another yttrium compound),
in an amount of 0.2 to 20% of the silicium nitride weight, the initial
silicium nitride containing also other components than yttrium. Isostatic compression of the composition is effected under a pressure of
100 MPa with a gas, acting as compression agent at a temperature of
1600°-1900°C.

In Patent (20/3) ASEA A.B. reveal a process for producing silicon nitride articles by the isostatic compression of pre-formed bodies or
powder, with a gaseous pressure agent (N), whereby the pre-formed body
(powder) is encased in a pulverulent layer, which is permeable to gases,
and which is transformed into a gas-impermeable casing, while the per-

meable pulverulent layer is in contact with a gas under pressure and
while a pressure is maintained in the gas that is at least as high as
the pressure prevailing at the same time in the pores of the powder
(pre-formed body). Such casing may be composed of glass of high smelting
point.

Patent (20/4) of ASEA A.B. (SE) refers to the production of articles of
silicium nitride (containing free silicium) by sintering, whereby the intervals existing between the particles of the initial material (powder)
will be filled up with nitrogen in such an amount that nitrogen can be
bonded with the silicium which is present in the initial material, whereafter the pre-formed body of powder is heated to effect the formation
of silicium nitride.

Patent (20/5) of ASEA A.B. describes the production of silicium nitride
articles, by isostatic pressure from a pre-formed body of SiC powder by
means of a gaseous pressure agent in a pressure vessel, the pre-formed
body being first placed in a glass casing, whereby at the end of pressing,
the pressure and temperature are so controlled that the pressure in the
vessel is reduced under the partial nitrogen pressure, at a temperature,
at which glass can be plastically deformed, thereby causing the expansion of the casing, so that the pre-formed body can easily be separated
from the glass surrounding it.

Another Patent (20/7) of ASEA A.B. refers to the production of ceramic
articles from, for example silicium nitride powder, which is mixed with
a softening agent and then formed into two parts, of which the final
product is then composed. The said parts may have different cross-
sections and consequently they are mixed with softening agents of cor-
respondingly different mol. weight. The parts are put together to obtain a composite product, which then is coated with glass and isostatically sintered.

Patent (21/2) of Association pour la Recherche et le Développement des
Méthodes et Processus Industriels (ARMINE) (FR) concerns a process for
producing articles of sintered silicium nitride, which contains aluminium (5-15 w.%), sintering being carried out in a nitrogen-rich atmosphere, which contains less than 1 vol.% of carbon monoxide. Due to

the oxidation by carbon monoxide of the silicium nitride forming in the
surface layers of the articles, a sufficiently open porosity is established therein, permitting the penetration of nitrogen into the interior
of the articles, realising a homogeneous nitriding thereof.

Patent (35) of The Boeing Company (US) describes a method for producing
pre-shaped alpha-silicon nitride whisker compacts and loose whiskers for
composite material reinforcement, by producing uncontaminated silicon
nitride whiskers, which previously could not be accomplished without
high cost. The present invention relates to the production of uncontaminated alpha-silicon nitride whiskers at a reasonable cost as either
shaped preform or individual whiskers by blending silicon particles
with composition particles; compacting the blend of silicon and composition particles to a shaped preform, subliming the composition particles
to produce a shaped porous silicon precursor, and reacting the precursor
with nitrogen. The shaped whisker compact can be impregnated with various
materials to form composites, or can be broken down to individual whiskers
to be used in molding powders.

According to Patent(71/1) of Daimler-Benz A.G. ceramic structural elements
can be bonded by a process, in which pre-sintered structural elements of
silicium nitride are introduced in a hot-pressing matrix. The sinterable silicium nitride, forming the elements, also contains small amounts
(up to 1% of the SiN powder weight) sintering adjuvants like magnesium
oxide. Prior to hot-pressing the pre-sintered elements are exposed to
an oxidation treatment.

Patent (71/3) of Daimler-Benz A.G. concerns the production of SiN articles, by shaping a plastic mass of silicium powder and an organic binder into the desired form, which then is dried on air or in an inert gas
atmosphere at $\pm$ 250°C into a green body, followed by hardening this body
and then by nitriding it in nitrogen atmosphere at up to 1400°C. The
binding agent consists of silicon resin, which remains in the green body
during hardening and which will be nitrided as well.

In Patent (71/5) of Daimler-Benz A.G. oxidation resistant silicium nitride sintered bodies of increased mechanical strength are described,

obtained by diffusing in the surface of the aluminium-nitride sintered
bodies aluminium oxide or aluminium nitride. Diffusion is effected by
calcining the silicium-nitride sintered bodies in aluminium oxide or
aluminium nitride powder at a temperature over $1300^\circ C$, during 3-40
hours.

Patent (73/1) of Degussa A.G. (DE) concerns a process for producing
ceramic structural elements of SiN and SiC by injection moulding through
mixing Si powder, SiN powder and SiC powder with thermo- or duroplastic
synthetic materials, burning out the synthetic material after shaping
the green body, sintering and converting SiN at high temperatures.
Prior to mixing with synthetic materials, the powder is treated with
silico-organic compositions.

Patent (76/2) of Deutsche Gold- und Silber-Scheideanstalt vormals Roesler
(DE) and Annawerk Keramische Betriebe GmbH (DE) describe a process for
producing silicium nitride in form of powder or shaped body by transforming elementary silicium with nitrogen or ammonium (at increased temperature) in the presence of a catalyst, which may be composed of compositions of phosphor and/or antimony, with nitrogen and/or silicium; or
of phosphides or antimonides. Volatile catalytic compositions of phosphor or antimony or phosphor vapours can be introduced in the nitriding
gas.

In Patent (76/3) the aforementioned applicants claim a process for
realising a strong bond between reaction-sintered and hot-pressed silicium-nitride structural elements by applying a silicium-nitride containing intermediary layer, consisting of a powder mix of silicium nitride
and aluminium oxide, the bonding zone being realised at a temperature
under $1600^\circ C$ in nitrogen atmosphere.

A process (76/5) developed by Deutsche Gold- und Silber-Scheideanstalt
is aimed at producing SiN composite bodies from reaction-sintered and
hot-pressed body parts through hot composite pressing at $1500^\circ - 1800^\circ C$
under a pressure of 15-30 MN/m^2 with the aid of a bonding layer consisting of a mixture of high smelting point oxides and silicium nitride.
The bonding layer contains 30-70% magnesium oxide, the rest being of

SiN. The powdery mixture of these components is applied to the surface, which will be the bonding surface, by flame-spraying.

Another Patent (76/6) of Deutsche Gold- und Silber-Scheideanstalt
refers to the production of shaped elements of silicium nitride. Part
of an element consists of 40-98% silicium of metallic or intermetallic
phases, like the silicide of magnesium, calcium, strontium, barium,
aluminium, copper, silver, zinc, tin and their mixtures or alloys. The
metallurgical process of forming the elements takes place simultaneously
with the bonding of various components into a composite element.

Patent (76/7) of Deutsche Gold- und Silber Scheideanstalt and Annawerke
also concerns a process for producing composite moulded articles of
reaction-sintered and hot-pressed silicium nitride. From the initial
material a blank is prepared by hot pressing, containing pressure adjuvants. The blank is placed between matrices of the same material,
which are adjusted to the shape of the blank and separated from one
another by an intermediary layer for preventing that the matrix parts
should be sintered together. Pressure is applied under such conditions,
which due to the suitable shape of the matrix parts and the predetermined orientation of the compression force, permit a selective compression of the blank.

Patent (81/3) of Dresser Industries, Inc. relates to a process for producing (in situ) nitride-bonded refractory bodies by mixing together
1-25 w.% silicium, 1-5 w.% raw clay, the rest consisting of a refractory material aggregate, the process comprising: the forming of the
mixture into a body, under pressure, sintering the body in a nitriding
atmosphere at high temperature. The aggregate is composed of a sintered
refractory clay, mullite smelt, synthetic aluminium oxide and magnesium
spinel.

According to Patent (95/1) of Fiat S.p.A. (IT) sintered silicon nitride
articles are prepared by sintering a silicon nitride compact, obtained
from silicon nitride powder, of a density of at least 1.3 g/cm^3 in nitrogen gas atmosphere at substantially atmospheric pressure, while the
compact is embedded in a protective powder comprising silicon nitride,
boron nitride, or a mixture of silicon nitride and boron nitride, and

one or more sinterisation additives for silicon nitride, said additives
being present in the protective powder in an amount of from 3 to 20%
by weight.

In another Patent (95/2) of Fiat a process is disclosed of producing
sintered silicon nitride articles from a silicon nitride compact by
forming a silicon powder, containing one or more sintering additives
for silicon nitride into a compact, having a density of at least
1.3 g/cm^3, the additives being chosen from MgO, Y_2O_3, CeO_2, ZrO_2, BeO,
Mg_3N_2, AlN, Mg_2Si, $MgAl_2O_4$, La_2O_3 and Fe, and being present in the
powder in an amount to ensure an additive content of from 0.5 to 20 w.%
in the silicon nitride compact, nitriding the silicon compact by heating
under nitrogen gas blanket at a temperature not exceeding 1500°C, to
convert the silicon into reaction bonded silicon nitride, and sintering
the reaction bonded silicon nitride compact by heating in a nitrogen
gas atmosphere at a temperature of at least 1500°C.

According to Patent (96/1) of Ford France S.A. (FR) silicium nitride
articles are formed from metallic silicium particles, the article thus
obtained being heated in a gaseous atmosphere (nitrogen, a mixture of
0.5-4.0% hydrogen and nitrogen), this nitrogen content being enough to
transform silicium into silicium nitride. The process yields articles
of improved resistance to traction and creeping forces.

Ford France S.A. developed also a series of silicium nitride articles,
consisting of elements of different density. Thus Patent (96/2) refers
to producing a silicium nitride article of triple density, by compacting
a mixture, which contains SiN and a densifying agent (0.5-5 w.%) under
heat and pressure, to form a first element of SiN. A second element is
formed by injection moulding of metallic Si particles and a support, the
second element being heated to burn out the support. A third element
is formed by injection of a mass containing metallic Si particles in a
carrier substance, the third element being heated in inert atmosphere
to increase the mechanical strength of this element, which then is placed
in the proximity of a part of the surface of the second element so that
a small space forms between them. The second element is bonded with
the third element by injection, whereby an injection mass containing
metallic Si-particles in suspension, is arranged in the space between the

two elements in order to accumulate metal particles in this space. The second and third element are bonded together in one operation in such a way that both elements are converted into pure SiN and joined together in the zone where the operation involving moulding has been effected.

Patent (96/3) of Ford France provides a process for producing a product of double density by forming a first element of a mixture of SiN particles and a densifying agent (MgO), the mixture being compacted by heat and pressure. A second element is then formed by injection of metallic Si, suspended in a carrier, the second element being sintered in inert atmosphere, followed by nitriding the second element in order to convert it into pure SiN, whereafter a surface of the first element is set into close contact with a surface of the second element in stabilised position, and both elements are heated, while pressure is only applied to one of the elements, thereby realising that a part of SiN, forming a part of a surface of one element, is bonded to the surface of the other element.

Patent (96/4) and essentially (96/2) of Ford France refer to a product of SiN of triple density, obtained by compacting a mixture of SiN particles and a densifying agent (5-0.5 w.%) and forming from this mixture under heating and pressure a first element of SiN; by forming through injection a second element of metallic Si particles and a support, the second element being heated to burn out the support; by forming a third element through pouring a mass in the proximity of at least one part of the second element, the poured mass containing Si metal particles, suspended in a carrier; the second and third elements being nitrided in one operation, thereby converting them into pure SiN, these two elements being bonded together in the zone of pouring. Thereafter a surface of the first element is set into a close contact with a surface of the third element, these elements being in stabilised position, while the adjoining surfaces of the first and third element are bonded by heating both elements and applying pressure to one of the elements, in the meantime the other element being held in stabilised state to realise that the SiN forming the surface of one of them forms a bond with the corresponding surface of the other element.

Patent (96/5) of Ford France concerns a SiN product having two distinct zones of different density by forming a porous body of metallic Si articles, the porous body being sintered in inert atmosphere; a casting being made of metallic Si and a carrier on at least a part of the surface of the said porous body, which absorbs the said carrier, promoting thereby the forming of a layer of metallic Si particles on the porous body's surface followed by nitriding simultaneously the metallic Si of the porous body and the deposit on the casting in an atmosphere of nitrogen gas, while the porous body is held at a temperature, capable of provoking a reaction between nitrogen gas and the Si metal body. In this mass a monolithic product is obtained, which displays different densities in the zones, formed by different processes.

A product of two different densities obtained by a process, described in Patent (96/6) of Ford France by compacting a mixture of silicium nitride particles and a densifying agent (MgO) to develop therein a density of 98% under heat and pressure, thereby obtaining a first element of SiN. A second element is formed by injection moulding silicium metal particles and a carrier substance in order to obtain a shaped second element, which is heat-treated to burn out the carrier and to convert the metallic silicium into pure SiN, whereafter a surface zone of the first element is set into contact with a surface zone of the second element and heat is applied to both elements, pressure only being applied to one element, thereby realising that a part of SiN forming the surface zone of one element is closely bonded with the surface zone of the other element. Bonding takes place at 1625°-$1700^{\circ}C$ under pressure of min. 70 kg/cm^2 during $1\frac{1}{2}$ hours. Thus, a product is obtained, consisting of one piece with parts of different density.

Ford Motor Company Ltd. (GB) developed a method (97/5) of treating a ceramic body of silicon nitride to protect the body from cracking during a hot pressing operation, in which pressure is applied to the body, the ceramic body having a portion of complex shape, defined by surfaces extending in different directions, the complex shaped portion of the body being supported by an encapsulating medium during the hot pressing operation. The method comprises the steps of applying a relatively thin layer of a release material (boron nitride) to the surfaces of the ceramic body, which are to be reaction surfaces during the hot pressing operation, this release material being non-reactive with the ceramic body and encapsulating medium at the temperatures and pressures encount-

ered in the hot pressing operation; and applying a relatively thick
layer of the release material to the surfaces of the ceramic body which
are not to be load reaction surfaces during the hot pressing operation.

Ford-Werke A.G. (DE) disclose in Patent (98/1) a process for producing
a silicium nitride article with a determined final density of 2.1 g/cm^3
- 3.18 g/cm^3, consisting of nearly pure Si metal in finely dispersed
state. 70-80 w.% of silicium metal particles and 30-20 w.% of a carrier
substance are used to suspend the metal particles in the carrier with
accurate adjustment of the suspension's viscosity and producing the end
product by top-pouring the Si metal particle suspension.

Another Patent (98/3) of Ford - Werke A.G. refers to the production
of silicium nitride bodies by preparing a mixture of Si-particles,
which, on reacting with nitrogen, will be transformed into silicium
nitride. The mixture is heated to a degree below the temperature at
which silicium particles and nitrogen react, whereafter the mixture is
heated in a nitrogen containing atmosphere, whereby the reaction zone
(which has already a temperature over the reaction temperature) is
moved from the surface of the body towards the interior thereof, so
that the body will be gradually nitrided starting from the body surface
towards its interior.

In Patent (98/4) Ford-Werke A.G. claim a process for producing a silicium nitride element, displaying a lower density than the theoretical
one and containing a compacting adjuvant. This element is coated with
a silicium nitride film, which is gas-impermeable and then heated to a
temperature, permitting the diffusion of a part of the compacting adjuvant into the SiN film, thereby improving its resistance against fissures.
Thereafter the element is exposed to pressure, permitting the increase
of the density of the SiN element to a value above its initial density
and pressing the SiN film into the SiN element, thereby forming a monolithic product.

According to Patent (104/13) of General Electric Company a body of polycrystalline silicium nitride is obtained by nitriding a sintered body
of polycrystalline silicium by forming a shaped body of silicium powder
(green body), which is sintered at a temperature between 1250°C and the

fusion point of silicium and displays a microstructure of particles and pores between 0.1 and 6 microns. The nitriding phase consists of a reaction of polycrystalline silicium in nitrogen gas atmosphere at a pressure, which is between subatmospheric and superatmospheric pressure at a temperature between 1100° and the fusion point of silicium, thereby obtaining polycrystalline silicium nitride.

Patent (104/17) of General Electric Company provides a process for obtaining a shaped polycrystalline body prepared from a homogeneous dispersion (with submicron granulometry) of silicium nitride in combination with a beryllium-base additive (e.g. carbide, fluoride, nitride of beryllium). From this dispersion a green body is formed, which then is sintered at $1900-2200^{\circ}C$ in a nitrogen containing atmosphere under a nitrogen pressure of 40 AT at $2000^{\circ}C$ and a nitrogen pressure of 75 AT at $2100^{\circ}C$. The nitrogen pressure established prevents the thermal decomposition of silicium nitride and produces a sintered product, having a mass volume corresponding to 80% of the theoretical mass volume of silicium nitride.

Another Patent (104/18) of General Electric Company reveals a process for producing a shaped polycrystalline body from a homogeneous dispersion of silicium nitride (with submicron granulometry), containing a magnesium-base additive (magnesium carbonate, magnesium nitride, magnesium cyanate, magnesium fluoride), furthermore a beryllium-base additive which can be a carbide, a fluoride, a nitride of beryllium or a nitride of beryllium and silicium. The dispersion is formed into a green body, which then is sintered at $1800^{\circ}-2200^{\circ}C$ under a nitrogen pressure of 20 AT at $1900^{\circ}C$ and of 40 AT at $2000^{\circ}C$. The final product displays a mass volume similarly to the product of Patent (104/17).

Patent (116/1) of GTE Laboratories, Inc. (US) refers to polycrystalline bodies, comprising a mixture of Si_3N_4, SiO_2 and yttrium oxide (Y_2O_3), the ratio between the number of silicium nitride moles and the number of moles of yttrium oxide being greater than 1. Yttrium oxide is present in the mixture in an amount of 3-13 w.%, while the concentration of SiO_2 is at least 3 w.%. The mixture also contains 0.1 w.% cationic impurities.

GTE Laboratories, Inc. developed in Patent (116/3) a densified abrasion resistant composite article, consisting essentially of particles of a hard refractory material selected from the group consisting of metal carbides, metal nitrides, or combinations thereof, uniformly distributed in a two-phase matrix, consisting essentially of a first phase and a second phase, the first phase consisting essentially of crystalline silicon nitride and said second phase comprising a refractory inter-granular phase of silicon nitride and a densification aid including magnesium oxide and silicon dioxide.

This process differs from that claimed in Patent (116/2) of GTE La-boratories, Inc. in that the hard refractory particles are selected from the group consisting of the carbides and nitrides of titanium, vanadium, chromium, zirconium, niobium, molybdenum, hafnium, tantalum, tungsten and combinations thereof.

According to a Patent (116/4) of GTE Laboratories, Inc. dense, poly-crystalline silicon nitride-based articles can be obtained by
a) providing a finely divided mixture of silicon nitride, silicon dioxide and yttrium oxide; b) pressing the mixture to form a sinter-able compact; c) embedding the sinterable compact in a non-sinterable setter bed powder mixture of about 48 mole % to 94 mole % silicon ni-tride, the balance consisting essentially of yttrium oxide and silicon dioxide, wherein the mole ratio of yttrium oxide to silicon dioxide is between zero and about 2; and d) heating the sinterable compact and non-sinterable setter bed powder mixture at a temperature and for a period sufficient to sinter the compact to a densified composite article having a density of at least 98% of theoretical.

Patent (116/5) of GTE Laboratories, Inc. provides sintered, polycrystal-line silicon nitride articles similar to those described in (116/4), in which the minor phase may consist essentially of compounds of silicon, yttrium, nitrogen and oxygen wherein the atom ratio of yttrium to sili-con in the minor phase of the surface layer portion of the article dif-fers from the atom ratio of yttrium to silicon in the minor phase of the interior body portion of the article.

Patent (116/6) of <u>GTE Laboratories, Inc.</u> also concerns a Si_3N_4 polycrystalline ceramic body, consisting essentially of a first phase of
Si_3N_4 and a second intergranular phase comprising SiO_2 and a densifying
additive, wherein the intergranular phase is substantially completely
crystalline. The Si_3N_4 body is prepared by sintering a body of particulate material, which is free of glass forming impurities or sintering
and heat treating a body of particulate material which contains glass
forming impurities.

The object of Patent (153) of <u>Kobe Steel Ltd. (JP)</u> is a process of
producing sintered silicium nitride (Si_3N_4) of high density (rel.density
98%), the pulverulent SiN being shaped and sintered with the addition
of a sinter adjuvant in form of an $Y_2O_3-Al_2O_3-MgO$ system. Sintering
takes place in two phases, the presintering phase compacting the body
to a 92% density, followed by hot pressing in an inert gas atmosphere
at $1500^\circ-2100^\circ C$ under a partial nitrogen gas pressure of min. 500 bar
(500 AT), until the required density of 98% is obtained.

Patent (170/1) of <u>Lucas Industries Ltd. (GB)</u> refers to a ceramic material
including at least 90 w.% of a single phase compound having a beta-phase
silicon nitride lattice, in which the silicon in the lattice has been
partially replaced by aluminium and the nitrogen has been partially replaced by oxygen, the degree of replacement being such that the amount
of aluminium and oxygen in the compound constitutes up to 75% by weight
of the compound.

Patent (170/8) of <u>Lucas Industries Ltd.</u> describes a method of forming
a sintered ceramic product from silicon nitride powder, wherein the
silicon nitride powder is heated at an elevated temperature with a first
metal oxide and a second metal oxide, the first and second metal oxides
being other than silica and being arranged so that during heating they
react with silica present as an impurity on the silicon nitride to form
a silicate having a liquidus temperature below that of the silicate,
which would be formed from silica with either the first metal oxide alone
or the second metal oxide alone. Preferably, the liquidus temperature
of the silicate formed from silica with the two metal oxides together
is at least $100^\circ C$ below that of the silicate which would be formed
from silica with either metal oxide alone and also, preferably, the
heating step is accompanied by pressure.

A method disclosed in Patent (185/2) of <u>Motoren- und Turbinen-Union</u>
<u>München GmbH (DE)</u> for preparing ceramic material bodies consists of
applying to the surface of the body a continuous porous layer of glass
and/or ceramic material and sintering to form a pressure-tight envelope. Sintering may be effected under a vacuum, but the body may
alternatively be evacuated prior to sintering and filled with nitrogen,
the latter embodiment being especially suitable for a body of reaction
sintered silicon nitride. Because by virtue of its porosity the layer
of glass and/or ceramic material can be applied prior to evacuation.
The encapsulating procedure can be considerably simplified. The continuous porous glass layer is formed from glass powder, containing pure
SiO_2 glass, and SiO_2-B_2O_3 glasses or glasses with additions of MgO, Al_2O_3.
The porous layer may consist of glass powder, containing lithium. Glass
powder can be applied in mixture with an organic binder (saturated alcoholic stearin solution or aqueous gelatine paste).

Patent (185/3) of <u>Motoren- und Turbinen Union München GmbH</u>, relates to
a method of manufacturing ceramic mouldings of silicon nitride, wherein
a blank is formed from a powdery mixture of silicon, alpha-silicon nitride
and a sintering agent together with an organic binding agent. The blank
is then heat-treated for removing the binding agent, nitrided to convert
the silicon to silicon nitride and subsequently subject to hot isostatic
pressing. Conventionally, the blank has been formed from silicon powder
only together with the binding agent with the result that the nitriding
step is very prolonged and the blank is extremely fragile so that during
handling prior to isostatic pressing there is a risk that it will disin-
tegrate. In the proposed method only the silicon powder component needs
to be nitrided so that the nitriding time is much reduced and, furhter the
alpha-silicon nitride component of the starting material hardens to produce a compact moulding which can readily be handled without risk of
disintegration. The sintering agent is one or more of MgO, Al_2O_3,
Y_2O_3, Cr_2O_3, Fe_2O_3, FeO and is present in an amount of (0.1 to 10w.%
of the powdery mixture and the blank is nitrided in a nitrogen atmosphere at 1100° to $1600^\circ C$.

Patent (185/7) of <u>Motoren- und Turbinen-Union München GmbH</u> also concerns the production of a cast body of silicium nitride and silicium
carbide by compressing the cast body, which is placed in a casing,

under vacuum (the casing being made of heat resistant glass). Compression takes place by applying pressure thereto from all sides and is carried out until all the pores disappeared in the body. The process is effected by arranging on the cast body the glass components in non-vitrified state, whereafter vacuum is established therein at the reaction temperature of the composition (800°-1500°C) until the non-vitrified components are transformed into a glass sheath, which envelopes the body and forms the casing.

In Patent (185/8) of <u>Motoren- und Turbinen-Union München GmbH</u> a process is presented for nitriding silicium bodies by supplying nitrogen between 1100°C and 1450°C, whereby the silicium body to be nitrided is so heated that first a thin layer of a surface of the body reaches the lower temperature limit, whereafter this layer is permitted to migrate towards the opposite surface of the body, nitrogen being continuously supplied from that side of the Si-body, which has not been nitrided as yet.

Patent (185/10) of <u>Motoren- und Turbinen-Union Münschen GmbH</u> describes a process for encapsulating a ceramic piece, more particularly a silicon ceramic piece, in order to subject it to a hot isostatic pressing without infiltration into the piece pores of the medium which transmits the pressure. The piece is previously evacuated and then filled with nitrogen (N_2) and finally immersed into melted silicon. The filling of the piece with nitrogen is carried out preferably under high pressure. After applying the silicon to the piece, an outside pressure can be exerted on the silicon.

Patent (187/2) of <u>National Aeronautics and Space Administration (NASA),US</u> reveals a metalceramic mass,of which a body is shaped containing particles of molybdenum, tungsten, tantalum and/or niobium, which are coated with a solid solution of silicium nitride, aluminium nitride, at least one of the oxides of aluminium and/or yttrium and/or chrome and dispersed therein.

<u>National Research Development Corp.</u> claim in Patent (189/1) a ceramic body of reaction-bonded silicium nitride, whereby at least the surface zone of the body is impregnated with an additive, consisting of calcium oxide, strontium oxide, a mixture of calcium oxide and strontium oxide,

with iron oxide or a mixture of magnesium oxide and aluminium oxide, (which contain the spinel $MgAl_2O_4$), these additives reducing the oxidation-caused strength loss of silicium nitride.

Patent (190/1) of NGK Insulators Ltd. concerns a sintered silicium nitride body, consisting of at least two metal oxides, selected from beryllium oxide, magnesium oxide and strontium oxide, each of these metal oxides being present in an amount of not more than 5 w.%, the rest being composed of silicium nitride. The metal oxide can be applied in combinations: beryllium oxide with magnesium oxide and/or strontium oxide. The body may also contain a rare earth metal oxide in an amount of max. 10 w.%.

Patent (193) of Nippon Carbon Co.Ltd. (JP) refers to shaped articles of an initial material, composed of carbon fibres, graphite powder, alumina powder, silicium nitride powder or silicium carbide powder. The production process includes a reaction between a mixture containing:
a) a component according to the formula:

$$R_{(4-n)} - Si - X_n$$

wherein: X = a halogen atom, n = a whole number between 4 and 1; R = a hydrogen atom, an alkyl group, an alkylene group, a vinyl group or an aryl group; b) a monomer, that can be copolymerised with a), furthermore glycerin or glycol, and c) ammonium gas or an ammonium generating compound.

The object of Patent (204/4) of Norton Company is a heat resistant high strength refractory composite, produced by sintering a mixture of silicon nitride, an yttrium compound and a thorium compound, the sintering refractory composite having a density of at least 65% of theoretical density, and the said mixture consisting essentially of from 80 to 98.5 w.% of silicon nitride, 1 to 15 w.% of yttrium oxide and 0.5 to 10 w.% of thorium oxide, with the total oxide present not exceeding about 20%.

In Patent (204/7) Norton Company claim monolithic, compact bodies of silicium nitride and/or silicium-aluminium oxynitride, containing also 10-50 % silicium carbide with a fracture modulus of at least 7000 kg/cm^2

at 20°C, for a 4-point contact on the ground and at least 2830 kg/cm^2 at 1375°C for a 3-point contact on the ground and a spec. resistance of 1 to 1 x 10^7 Ω .cm. The silicium nitride component mainly consists of the beta-phase; the silicium nitride component displays an expanded beta-silicium nitride structure.

Patent (204/9) of Norton Company concerns refractory silicium nitride bodies of high strength, composed of 0.5-15 w.% cerium oxide and 99.5-85 w.% silicium nitride with a density of 85% of the theoretical value. The silicium nitride body is obtained by hot pressing the powdery mixture of silicium nitride (average particle size: < 9 μm) and cerium oxide (particle size: max. 10 μm) at 1680-1800°C under a pressure of 140-840 kg/cm^2 during 10 minutes to 2 hours in a non-oxidising (inert) atmosphere.

According to Patent (204/15), Norton Company developed a process for intensifying the resistance of sintered silicium nitride by impregnating the surface of the shaped silicium product with aluminium oxide, followed by sintering in nitrogen atmosphere at 1200°-1500°C, while maintaining a considerable partial pressure of SiC in the atmosphere during sintering.

Patent (205/1) of Novatome (FR) describes a refractory material, consisting of a cement-bonded silicium nitride aggregate, which forms 50-90% of the final product's weight and is obtained by reaction-sintering. The cement bonding agent is composed of calcium aluminate. Silicium nitride and calcium aluminate are first mixed in dry state, then blended in a humide mixer with a water content of 15%.

Patent (209/1) of The Plessey Co.Ltd (GB) reveals a method of producing a silicon nitride article, which includes the steps of providing a dense silicon nitride compact, forming the compact into a desired shape, providing a silicon powder, mixing the silicon powder with a binding agent, forming the silicon powder mixture around the dense silicon nitride compact to form a silicon powder compacted layer around the dense silicon nitride compact, and at least partially nitriding the silicon powder compacted layer. The compact is formed into a desired shape by hot forging and the silicon powder has a powder particle size of not greater than 200 British Standard Mesh. The silicon powder mix-

ture contains about 15 w.% of the binding agent and the silicon powder
mixture is pressed around the hot forged dense silicon nitride compact
at a pressure of about 25 tons per square inch.

Another Patent (210) of Plessey Inc. refers to a method of manufacturing laminae of dense silicon nitride, which includes a first stage in
which silicon nitride powder is mixed with a binder to form a slip;
a second stage in which the slip is cast as a film on a smooth surface
and allowed to dry; a third stage in which sheets of the cast film are
arranged in parallel layers to form a stack in which at least some of
the said sheets are separated from one another, the separation being
effected by boron nitride; a fourth stage in which the silicon nitride
in the sheets is made dense by treating the stack with heat and pressure;
and a fifth stage in which the treated stack is dismantled to provide
independent laminae of dense silicon nitride, composed of at least one
of the said sheets. The layers of boron nitride for effecting the separation of at least some of the said sheets, are formed by mixing boron
nitride powder with a binder to form a slip; casting the slip as a
film on a smooth surface, and allowing the film to dry. The binder
applied includes a lacquer, a plasticiser and a release agent. The
lacquer comprises: 360 mg of cellulose-acetate butyrate; 1500 cc of
n-propyl acetate; 1500 cc of methyl ethyl ketone; 450 cc of mesityl
oxide; 375 cc of cyclohexanone, and 225 cc of diacetone alcohol.
The patent simplifies the machining and grinding operations of thin
laminae with a thickness between 0.002 and 0.02 inches.

Patent (211) of R.Pompe (SE) provides a process for the production of
moulded bodies, consisting of materials based on silicon nitride,
starting from mixes of 40-48 w.% silicon powder (Si), silicon nitride
powder (Si_3N_4) and a sintering aid, entailing the preparation of a powder mix, moulding according to known methods, nitriding of the Si-content
to Si_3N_4 and sintering. The used silicon powder is crushed to a mean
particle size below 1/um in the presence of silicon nitride powder as a
dispersing agent and possible sintering aids and, a body moulded of this
powder mix is then nitrided during a time of   not longer than 5 hours,
whereby the nitriding temperature is kept below the melting point of pure
silicon.

According to Patent (214/7) of The Research Institute for Iron, Steel and other Metals of the Tohoky University shaped, sintered articles are produced from a mixture of a powder of metal nitride with a binder, selected from the group containing: compositions with only Si-C bonds; compositions with SiH + SiC bonds; compositions with SiH Al bonds; compositions with Si-Si bonds; compositions with Si-N bonds, and silicium compositions of high mol. weight, wherein silicium and carbon are the main components. The free C still present in the article is removed by heating to 800°-1400°C in oxidising atmosphere. The bonding agent is applied in an amount of 0.3-45 w.%. The mixture of metal nitride powder and the binder is shaped by hot pressing in one cycle, simultaneously with sintering.

Patent (220/1) of Rolls-Royce Ltd. (GB) presents an article made of a silicon nitride or silicon carbide powder material by a method, comprising: providing, in a die, powder material, having a non-uniform dimension in the direction of pressing and applying pressure to the powder material to make a pre-form of the article to be manufactured, the pressure being applied initially to that part of the powder material which has the greatest dimension in the direction of pressing and then to the remainder of the material in order of diminishing dimensions in the direction of pressing. The powder material comprises silicon nitride and a densifying agent, which comprises magnesium oxide in the ratio one part magnesium oxide to twenty parts silicon nitride (w.%). Hot pressing takes place at a pressure of about 3000 p.s.i. and a temperature of about 1850°C.

Patent (221/3) of Rosenthal A.G. (DE) refers to a process for reducing the gas permeability of porous bodies, composed of reaction-sintered silicium nitride, through impregnation and firing in nitrigen atmosphere. More particularly the body is impregnated, for example, with $SiCl_4$ or $SiBr_4$. Under the effect of a treatment with ammonium, the silicium imide, forming by a conversion with the silicium halogenide, is precipitated in the cells and then, through firing in nitrogen atmosphere, is converted into silicium nitride, filling up the pores.

Patent (246/5) of Sumitomo Electric Industries Ltd. relates to a method of sintering silicon nitride for obtaining a silicon nitride sintered

compact, having advanced high temperature strength and high density. In
this method silicon nitride powder, mixed with at least one kind selected from among the metallic oxides except silicon oxide and/or at least
one kind selected from among the nitrides and carbides of the IVa, Va,
VIa metals, is pressed and sintered. The heating process is effected
in atmospheres of different conditions varying by stages from the ele-
vation of temperature to the completion of sintering, i.e. a vacuum below 10^{-3} atm. for the first stage, a partial pressure atmosphere of
10^{-3}-0.9 atm. for the second stage, and a high pressure atmosphere
above 1 atm. for the third stage. The method enables to impart the
aforedescribed characteristics to a silicon nitride sintered compact,
although it has been considered difficult to sinter, due to its covalent
bending, e.g. Si_3N_4 powder alone, and to produce a sintered compact of
advanced high temperature strength and high density even when sintered
in a vacuum, under normal pressure or under pressure at a high temperature by using a sintering additive such as metallic oxides, nitrides,
carbides, etc.

Patent (254/8) of Tokyo Shibaura Denki K.K. (JP) describes a process for
preparing a powdery silicon nitride composition, which comprises: adding
to silicon, having a grain size of ca. 0.01 to 0.05 um (a) carbon having
a grain size of ca. 0.01 to 0.05 um and (b) one or more metallic compound(s),
selected from (1) the group consisting of salts or oxides of Mg, Ca, Sr
and Ba, (2) Al, Al_2O_3, AlF, AlN, Fe_2O_3, SnO_2, Nb_2O_5, La_2O_3, Cr_2O_3,
CeO_2 and Cr_3C_2 reducing and nitriding the resultant mixture. The process
for preparing a sintered body comprises: moulding the thus produced
composition in the presence of a sintering additive into a desired shape
and then sintering the shaped body at a temperature of ca. 1600 to $1800^{\circ}C$.
The sintered body exhibits excellent strength and can be produced from
the said composition with a shortened reaction time.

The object of Patent (254/9) of Tokyo Shibaura Denki K.K. is a process,
which comprises: mixing powder (A) of silicon nitride powder and powder
(B), obtained by heat-treating a powdery mixture of silicon nitride powder and a sintering additive in a non-oxidising atmosphere and then grinding the resulting heat-treated products into powder; and forming the
resultant powdery mixture into a desired shape, which is then sintered
in a non-oxidising atmosphere. The powder (B) is heat-treated at tempe-

ratures of 1450-1750°C and consists of 40 to 60 w.% of the sintering additive and the balance of silicon nitride powder. The sintering additive
comprises one or more compounds selected from the group consisting
of MgO, Al_2O_3, Y_2O_3, SiO_2, TiO_2, ZrO_2, Li_2O, CaO, BeO, V_2O_5, MnO_2,
MoC_3, WO_3, Cr_2O_3, NiO, Nb_2O_5, Ta_2O_5, HfO_2, oxides of other rare-earth
elements, AlN, TiN and SiC.

Patent (255/3) of Tokyo Shibaura Electric Co.Ltd. (JP) describes the
production of alpha-silicium nitride powder, obtained by mixing together
silicium dioxide powder, carbon powder, metallic Si-powder, the mixture
being heated in a nitrogen or ammonium containing atmosphere to the
nitriding temperature, thereby obtaining an SiN product, which then is
heated in an oxidising atmosphere to separating carbon therefrom. The
powder consists of 1 w.% silicium dioxide powder, 0.4-4 w.% of carbon
powder and 0.1-2 w.% metallic Si powder.

In Patent (255/5) of Tokyo Shibaura Electric Co.Ltd.also developed
a pulverulent ceramic material of silicium nitride, wherein the amount
of oxygen combined with inevitable impurities, is less then 2 w.%. The
pulverulent material contains,in addition to silicium nitride,0.05 -
2.5 w.% aluminium and 0.4 - 8.0 w.% yttrium.

According to another Patent (255/6) of Tokyo Shibaura Electric Co.Ltd.
alpha-silicium nitride powder is obtained by preparing a mixture of
silicium dioxide powder, carbon powder and silicium nitride powder or
silicium carbide powder or silicium oxynitride powder, the mixture being
composed in the following ratio: 1:0.4 to 4:0.005 to 1.0. The mixture
is heated in non-oxidising atmosphere, containing nitrogen or ammonia,
the subsequent reduction and nitriding reactions (at 1350° - 1550°C)
yielding the end product: silicium nitride.

Toshiba Ceramics Co.Ltd. (JP) disclose in Patent (256/1) a silicium
nitride powder with a high (75 w.%) content of the alpha-phase and a
nitrogen content of at least 33 w.%. The silicium nitride particles hae
an average max. size of 3 μm, the major part of the powder particles
displaying the form of fibres or needles with a thickness of at most
1 μm. The starting material consists of silicium oxide powder and

amorphous carbon powder, having an oil absorption capacity of at least
100 ml/100 g. This powder mixture is then nitrided and fired at a temperature of 1300°-1700°C, under a nitrogen containing, non-oxidising
gas atmosphere. The weight ratio for carbon powder to silicium oxide
powder amounts to 1-3.

Patent (258) of K.K. Toyota Chuo Kenkyusho (JP) relates to a process
for producing sintered bodies, mainly of silicium nitride by mixing together 65-96 mol.% silicium nitride powder with 35-4 mol.% metal oxide
powder, for example an oxide of yttrium (Y_2O_3) of cerium (CeO_2), in
mol. ratios between 3:7 to 7.3. This powder mixture is shaped into a
green body in a non-oxidising atmosphere without pressure.

In Patent (267/2) UKAEA developed a process for producing an electrically conducting article, which is bonded by silicon nitride and the
electrical conductivity is provided by an electrically conducting phase
of silicon or of silicon carbide. A silicon nitride article is heated
to a temperature high enough to decompose part of the silicon nitride
to silicon, which is formed, the heating taking place either in an environment which is inert to silicon or in a carbon-containing environment
which reacts with the silicon formed to give silicon carbide. The temperature is in the range from 1350°C to 1550°C and the environment which
is inert to silicon comprises an atmosphere of argon or of helium or
vacuum.

Patent (267/4) of UKAEA refers to a method of preparing silicon nitride
articles, the method comprising the step of nitriding, to form silicon
nitride, a green body comprising silicon and an additive, wherein the
additive is boron or compound of boron (boron nitride), which is present
in the green body in a proportion in the range of 1% to 5 w.%. The
silicon in the green body is in the form of a powder having an average
particle size of less than 25 microns (5 microns) and the additive is
in the form of a fine powder having a surface area in the range from
5000 to 20,000 cm^2/g. It was found that the rate of   nitriding in
the present method may be decreased in a controlled manner compared with
the rate of nitriding of silicon alone. This is useful in the nitriding
of green bodies comprising silicon, where unavoidable temperature dif-
ferences may arise within the artefact, thereby giving rise to non-uniform

nitriding. Thus, different parts of a green body may be provided with different concentrations of boron or compounds of boron to obtain a uniform nitriding rate.

Patent (267/5) of UKAEA provides a method of joining a first surface, consisting essentially of non-fibrous silicon nitride to a second surface consisting essentially of non-fibrous silicon nitride, which comprises the steps of providing, interposed between the surfaces to be joined, a mixture comprising silicon, a binder, and a solvent for the binder; removing the binder, and firing in a nitriding atmosphere to convert the silicon to silicon nitride. As the binder polyvinyl butyral, while as the solvent for the binder methyl ethyl ketone may be used. The mixture may also contain a plasticiser, for example dibutyl phthalate. The first and second surfaces may, for example, define a crack in a cracked silicon nitride body. The method may, therefore, be applied to the repairing of such cracked articles.

A method claimed in Patent (267/6) of UKAEA relates to preparing a silicon nitride body, by (i) heating silicon and an additive comprising a Group IIA metal or a compound of a Group IIA metal in an atmosphere which is inert to the silicon and to the additive; (ii) forming the product of step (i) into a green body and (iii) nitriding to form the said silicon nitride body. As an additive, magnesium can be used. Silicon should be applied in the form of a fine powder, for example with an average particle size of 5 microns. The additive should be used in form of a fine powder as well, with a surface area of 5,000 to 20,000 cm^2/g.

Patent (267/7) of UKAEA concerns a method of making a refractory product, for example for a heat exchanger, having predetermined internal structural features, which comprises: adhering a plurality of components comprising refractory material in powder form and a binder, one or more of the components, having apertures to cooperate to define the said features when they are adhered; and subsequently heat treating the adhered components to give the artefact. The refractory material is silicon and the heat treating is carried out in a nitriding atmosphere to convert the silicon to silicon nitride. The components are adhered by applying an adhesive to those surfaces of the components, which are to confront one another when they are adhered, laying-up the components, and applying pressure thereto. The adhesive is a solvent for the binder and may be ethanol, toluene, polyvinylbutyral.

In Patent (267/8) UKAEA disclose a method of fabricating a shaped refractory body which comprises (i) shaping a mixture comprising a refractory material and an organic binder (SiN, polyvinylbutyral) in a mould by compression moulding at elevated temperature, wherein the mould includes a pre-formed rigid pattern comprising a resin, which may be a cold-setting resin, for example a filled epoxy resin, the filler being alumina; (ii) removing the shaped mixture produced in step (i) from the mould, and (iii) subsequently heating the shaped mixture so as to remove the binder, and subsequently firing to form the refractory body. The preformed rigid pattern is provided with a coating of a bismuth-tin alloy.

Patent (267/9) of UKAEA describes a method of preparing porous silicon nitride which comprises the steps of (i) preparing a green mixture comprising silicon powder, a binder and an inert, removable additive powder, wherein the additive has a particle size of less than 25 microns; (ii) curing the binder in the mixture; (iii) heating so as to remove the binder and (iv) firing in a nitriding atmosphere to convert the silicon to silicon nitride and thereby produce the porous silicon nitride. The additive is removed between steps (ii) and (iv) by being convertible to gaseous products. As an additive cellulose or amorphous carbon or graphite and as a binder polyvinyl butyral can be used.

Patent (267/10) of UKAEA relates to a unitary, cellular, silicon nitride body, having a first cellular portion, comprising silicon nitride, adhered to a second cellular portion comprising silicon nitride, wherein the cells of the first portion are different in size and/or type from the cells of the second portion. The cells of the first portion comprise open pores and the cells of the second portion comprise enclosed cells. The method of preparing a unitary, cellular silicon nitridy body comprises the steps of (a) adhering a first cellular body comprising silicon or silicon nitride to a second cellular body comprising silicon or silicon nitride by means of a mixture comprising silicon and a binder (polyvinyl butyral), (b) curing the binder in the mixture, and (c) debonding so as to remove the binder and firing in a nitriding atmosphere to produce the unitary, cellular silicon nitride body.

The object of Patent (267/11) of UKAEA is the production of a foamed, inorganic refractory material, obtained by preparing a mixture of powdered inorganic refractory material, and a binder under conditions such that gas or vapour is entrapped within the mixture (i); and (ii) subjecting the mixture to subatmospheric pressure to foam the mixture thereby to produce a green foam; and (iii) subsequently debonding to remove the binder and firing to produce the final foamed, inorganic refractory material. In step (i), the refractory material and the binder are mixed and then mechanically agitated in air so that the air is entrapped within the mixture. The refractory material in step (i) is silicon and the firing is carried out in a nitriding atmosphere to convert the silicon to silicon nitride which constitutes the final refractory material. The sub-atmospheric pressure is in the range from 30 mm Hg to 60 mm Hg which is maintained while the binder sets, or is set. As binders, cold-setting resins, such as epoxy, phenolic, polyester, silicon and polyamide resins can be used.

In Patent (267/12) UKAEA provide the protection of silicon nitride products, more particularly a product comprising SiN can be protected from molten aluminium attack by an adherent surface coating, comprising an oxide, other than aluminium oxide, whose thermodynamic stability is substantially equal to or is greater than that of aluminium oxide. The product contains between 0.5 and 5 w.% of oxide based on the weight of silicon nitride. The oxide may be a single oxide or may be a combination of one or more oxides. Examples of suitable oxides are oxides of elements of Group IIa of the Periodic Table such as BeO, MgO, CaO, SrO and BaO; oxides of elements of Group IIIa such as Y_2O_3 and La_2O_3; oxides of elements of Group IVa such as TiO_2, ZrO and HfO_2; oxides of the lanthanides such as CeO_2; and oxides of the actinides such as ThO_2 and UO_2. It is recommended to use BeO, MgO and BaO. The method of making such products comprises: providing a surface layer of a compound which is decomposable to an oxide, other than aluminium oxide, whose thermodynamic stability is substantially equal to or is greater than that of aluminium oxide, on a product comprising silicon nitride, and decomposing the compound to the oxide under conditions to form the ad-

herent surface coating. The surface layer is provided by immersing the
product comprising silicon nitride in an aqueous solution of the compound,
where the compound is water-soluble, followed by drying. The compound
is a water-soluble nitrate and is decomposed by calcining at a temperature
of less than 1000°C.

In Patent (267/13) UKAEA developed a method of preparing a foamed, refractory material comprising the steps of (i) preparing a mixture
comprising a material which is refractory or which is converted to a
refractory material in the firing step (iv) below, precursors which
react to give a foamable binder and a blowing agent for the binder; (ii)
subjecting the mixture to supra-atmospheric pressure whilst the mixture
is foaming; (iii) allowing the mixture to set, whilst maintaining the
supra-atmospheric pressure, to give a green foam, and (iv) debonding
to remove the binder and firing to produce the final foamed, refractory
material, the steps (ii) and (iii) being carried out in a mould and the
supra-atmospheric pressure in step (ii) is up to 50 p.s.i. in excess of
atmospheric pressure. The initial material is silicon and firing is
carried out in a nitriding atmosphere to obtain foamed SiN. The precursors used can be of the kind which, while reacting, develop a foamable
polyurethane.

Patent (267/20) of UKAEA reveals a process for producing an electrically conductive material by converting a mixture of silicium and another
component, under nitriding conditions, into an electrically conductive
phase, while forming a material (article, body) that contains silicium
nitride and the electrically conductive phase. Nitriding takes place
in an atmosphere, which does not oxidise the mixture, (e.g. nitrogen).

According to Patent (268) granted to the United States of America,
Department of Commerce (US), a porous reaction-sintered silicon nitride
body is infiltrated with an organosilicon compound after which the body
is heated at a temperature sufficient to decompose the infiltrated
material, resulting in a silicon nitride body having an increased density
and significantly improved room temperature strength.

10.3   Other silicates containing intermediates

<u>Nobutoshi Daimon, et al. (JP)</u> (202/2) developed a synthetic tetra-
silicic mica, which can be cleft into ultrafine particles by hydration
and which has a structure according to the following formula:

$$Na_{0.6-0.8}Mg_{2.6-2.7}(Si_4O_{10})F_2$$

The ultra-fine mica particles of this invention can be formed into
electrically insulating film, heat-resistant sheet and a composite with
synthetic resin. These products have excellent heat-resistance and electrically insulating properties. They can also be used in combination
with mineral fibers such as glass, silica, alumina, or silicates to prepare a non-flammable sheet, which is highly flexible. The flexibility
is due to the fact that the ultra-fine particles of mica are flake-like
particles uniformly cleft to molecular size, and the fact that the
flakes are reformed into a product by uniform overlapping. In addition
to the above uses, this synthetic mica can be used as a base for various
paints and as a starting material for the preparation of non-flammable
building materials.

Patent (219/3) of <u>Rhône-Poulenc Industries (FR)</u> concerns a process for
preparing silicium oxide and metal silicates, whereby an aqueous solution of an alkaline silicate is treated with an organic polar liquid
(for example methanol), that can be mixed with water. The metal silicates can be used in glass production.

Patent (275/1) of <u>Vereinigte Grossalmeroder Thonwerke (DE)</u> relates to
elements, promoting sintering,of increased resistance to temperature
changes, mainly in form of a plate for sintering a green body in a high
speed sintering oven, the plate being made of an initial material like
chamotte, cordierite, sillimanite, SiC or their combination, furthermore an additive like quartz glass, forming 5-50% of the total composition.

<u>10.3.1</u> Si_3N_4 containing intermediates

<u>Annawerk Keramische Betriebe GmbH (DE)</u>   reveal in Patent (16/2) a
process for producing shaped bodies of silicium nitride (Si_3N_4) according to which a pressing adjuvant (oxide) is introduced in these parts
of the body, which have to be compacted, hot pressing being carried out
at 1600°-1800^oC and under pressures between 150 and 500 kp/cm^2 by means
of profiled press dies.   At the beginning of the process, the initial
silicium carbide material is impregnated (under vacuum) with a solution
(smelt), which will form in a later phase, the pressing adjuvant (oxide)
by thermal decomposition.

Another Patent (16/3) of <u>Annawerk Keramische Betriebe</u> discloses a process
for producing refractory shaped bodies of silicium nitride or composite
silicium nitride materials by bonding a secondary part to a hot-pressed
primary part of the body, the secondary part consisting of a mixture of
50 w.% metallic silicium powder, a temporary binder and (as the case may
be) one or more of the compositions: Si_3N_4, SiC, Al_2O_3, the mixture
being bonded to the primary part, followed by the removal of the binder
by heating in nitrogen atmosphere, whereby silicium is converted into
silicium nitride, thereby realising the tight bond between the two parts.

The object of Patent (27/1) of <u>Battelle Memorial Institute (CH)</u> is a
silicon nitride-based powder composition, enabling materials of densities exceeding 3.10 g/cm^3 to be obtained by pressureless hot sintering.
The composition contains, intimately mixed by pulverisation, Si_3N_4
having a particle size not exceeding 1/um, and up to 6 w.% of a densification aid, comprising very finely ground magnesium oxide and aluminium oxide in a weight ratio from 10:1 to 1:3. This composition enables mechanical parts to be economically manufactured by moulding,
followed by sintering, their properties being practically equivalent to
those of parts obtained by machining hot pressed blocks of Si_3N_4.

<u>Elektroschmelzwerke Kempten GmbH</u> reveal in Patent (90/6) a practically
pore-free shaped body of polycrystalline silicium nitride and silicium
carbide in the form of a homogeneous micro-structure (particle size:
max. 10/um), with a weight ratio between pure Si_3N_4 and SiC powder,

varying between 5:95 and 95:5, without the addition of sintering adjuvants, through isostatic hot pressing in a vacuum-proof casing at
1800°-2200°C under a pressure between 100 and 400 MPa.

Patent (107) Gesellschaft für Kernforschung mbH (DE) describes a composition on the basis of Si_3O_4, which is coated with a protective coating
and is suitable for moulded ceramic products or structural elements of
high deformation resistance. The Si_3N_4 (commercial quality) composition
is provided with a compact coating also composed of Si_3N_4 or SiC. The
coating thickness is between 10 μm and 100 μm.

In a process developed by GTE Laboratories, Inc. (116/7) the addition
of controlled amounts of Al_2O_3 to high purity Si_3N_4 powder (containing
less than 0.1 w.% cation impurities and containing Y_2O_3 as a densifying
additive) enables shorter sintering times to achieve polycrystalline
Si_3N_4 bodies, having densities approaching theoretical density, while
a postsintering crystallisation heat treatment results in strengths at
high temperatures not otherwise obtainable in the presence of Al_2O_3.
The resulting Si_3N_4 bodies are useful as engine parts and components
of regenerator or recuperator structures for waste heat recovery.
In addition, such Si_3N_4 bodies containing Al_2O_3 exhibit good oxidation
resistance.

Patent (117) granted to GTE Products Corp. (US) describes a process for
producing a monolithic Si_3N_4 ceramic body, whereby at least a portion
of the body has an outer layer exhibiting a lower density than the interior thereof. The body contains up to 25 w.% of a densifying additive,
consisting of MgO, CrN, Y_2O_3, La_2O_3, ZrO_2, ZrN, HfO_2, CeO_2, Al_2O_3 or
SiO_2 and is produced by one-step sintering of green compacts in an ambient atmosphere of water vapour and a non-reactive diluent gas (e.g.
hydrogen or a noble gas), sintering taking place at a temperature between
1500°C and 1825°C. The ceramic articles obtained are useful in
special structural applications, such as abradable seals for gas
turbine engines and impact resistant surfaces.

In Patent (151) Kernforschungszentrum Karlsruhe GmbH (DE) reveal the
production of shaped SiN-bodies, obtained by pseudo-isostatic hot
pressing; the matrix of the hot press oven consists of carbon. The

initial Si_3N_4 material is treated in vessels, the lining (body) and the grinding elements of which are made of the same material (Si_3N_4). As a grinding fluid an organic composition like acetone and cyclohexane are used.

Patent (156) of M.N.Korotkevich, et al. (USSR) concerns a method of preparing silicon nitrides, oxides or oxynitrides of refractory metals and solid solutions thereof by evaporating a metal or an alloy thereof in the form of a consumable electrode in an electric arc, in a gaseous medium of nitrogen and/or oxygen at a pressure of from 2 to 7 atmosphere, wherein the arc is an alternating current arc, operating at a current of 5 to 15 amps per cm^2 of the electrode working surface, an electrode voltage of from 10 to 20 kW and an arc length of from 10 to 15 mm. The nitrogen and/or oxygen is diluted with a gas, which is inert both to the starting metals and to the final product, for example argon. According to the inventors no industrial methods of producing oxynitrides as individual compounds are described in literature, and they claim that the proposed method is simpler than the known ones, not requiring prolonged sintering of the initial mixtures at high temperature.

According to Patent (162/1) of Kurosaki Refractories Ltd. (JP) $SiC-Si_3N_4$ composite system is produced by firing, in a nitriding gas atmosphere, a green compact prepared from and composed of, as starting materials, silicon powder and an organic silicon polymer containing carbon and silicon atoms as the major skeletal components, whereby the composite system, as a final fired compact, has an interwoven texture of SiC and Si_3N_4 with sufficient micro-gaps to absorb thermal stresses, the quantitative ratio by weight of SiC to Si_3N_4 in the composite system being in the range of 5% - 20% : 95% - 80%.

The object of Patent (176/8) of Max Planck Gesellschaft zur Förderung der Wissenschaften e.V.   is a process for the sintering of Si_3N_4-cast bodies, by crushing Si_3N_4 with crushing elements, consisting of an oxide or a silicate of aluminium, zirconium, magnesium, glucinium and/or yttrium or Si_3N_4, with a specific surface of 10.5 - 35 m^2/g and a particle size between 0.2 - 0.05 microns. The obtained powder is shaped and sintered in an inert or nitrogen atmosphere at $1700°-1900°C$.

Patent (204/10) of Norton Co. (US) describes refractory articles of high
cold and hot stability and creep resistance, composed of 80-98% Si_3N_4,
1-15% Y_2O_3, 0.5-10% ThO_2, the sum of oxides being under 20%. The articles
display a density of 65-98% of the theoretical value, the deviation from
linearity $< 15\%$ and transverse strength being > 7000 kg/cm^2 at room
temperature and > 4900 kg/cm^2 at 1375°C.

Patent (267/3) of UKAEA provides a method of fabricating a support member on a refractory body, which comprises the steps of: adhering a support member to the body by means of a green mix, the support member comprising Si_3N_4 or Si_3N_4 and BN, protected from oxidation by a coating
comprising SiO_2 and B_2O_3, and the mix being one which is convertible,
by heating, to a product comprising Si_3N_4 or Si_3N_4 and BN protected from
oxidation by a coating comprising SiO_2 and B_2O_3; and heating to convert
the green mix to 1150° - 1350°C into a product comprising Si_3N_4 or
Si_3N_4 and BN protected from oxidation by a coating comprising SiO_2 and
B_2O_3. The product therein comprises a catalyst support having a honeycomb structure. The green mix additionally comprises a binder, for
example polyvinylbutyral.

10.3.2 Silanes

Patent (79/1) of the Dow Corning Corporation (US) refers to silicon
carbide prepolymers, which exhibit good handling properties, and are
useful for preparing ceramics, silicon carbide ceramic materials and
articles containing silicon carbide. Such prepolymers are polysilanes,
consisting of 0-60 mole% $(CH_3)_2Si$ units and 40-100 mole% CH_3Si units,
all Si valences being satisfied by CH_3 groups, other Si atoms, or by H
atoms, the latter amounting to 0.3-2.1 w.% of the polysilane. They
are prepared by reducing the corresponding chloro- or bromo-polysilanes
with at least the stoichiometric amount of a reducing agent, e.g.$LiAlH_4$.

Various types of ceramic materials have been developed by Lucas Industries
(GB). Patent (170/2) granted to this firm reveals a method of producing
a ceramic material, containing at least 90 w.% of a single phase silicon
aluminium oxynitride, comprising:mixing not more than 75% by weight of
alumina in powder form of particle size less than 10 microns, or a compound of aluminium which decomposes to give the required alumina at the

elevated temperature of the process, with powdered silicon nitride of
particle size less than 20 microns, and sintering the mixture at a temperature in the range 1550°C to 2000°C for at least 30 minutes to produce the ceramic material. The reaction materials are surrounded by a
protecting medium while at the said elevated temperature using as a protecting medium boron nitride powder.

A method, claimed in Lucas-Patent (170/10) consists of joining metal and/or
metal alloy components by electric welding, wherein the components are
retained in a required relative orientation during welding by locating
means which is engaged with the components and is formed, at least at the
region thereof presented to the components, of sintered silicon nitride
or a sintered ceramic product containing at least 80% by volume of a
substituted silicon nitride. Preferably, said sintered silicon nitride
or said sintered ceramic product is arranged to have a mean modulus of
rupture value at 25°C of at least 90,000 p.s.i. The substituted silicon
nitride is a single phase silicon aluminium oxynitride according to the
formula:

$$Si_{6-z}Al_zO_zN_{8-z}$$

where z is greater than zero and less than or equal to 5. It is recommended to carry out the joining operation by resistance welding.

Patent (170/11) of Lucas Industries discloses the production process of
a sintered ceramic material, which contains more than 95 w.% of a composition corresponding to the formula:

$$Si_{6-z}Al_zN_{8-z}O_z$$

(wherein z = 0-5). More particularly, the ceramic material is obtained
by heating to 1200°-2000°C a mixture of 15-45 w.% silicium oxide,
0.05-50 w.% of aluminium oxide, 40-60 w.% of aluminium nitride, the mixture also containing two metal oxides other than silicium oxide, which
act on a part of silicium oxide in such a way that in the mixture a
silicate glass is forming, this glass promoting the densification of
the ceramic material.

Patent (170/12) of Lucas Industries relates to a process for producing
a ceramic material by heating to 1200°-2000°C a mixture, which contains

silicium nitride, aluminium nitride, aluminium oxide and silicium oxide, and wherein a part of silicium oxide is present in the form of an impurity containing silicium nitride and aluminium nitride. The relative proportions of the aforementioned components in the mixture are similar to those given in Patent (170/11).

Patents (170/13) and (170/14) of Lucas Industries Ltd. refer to a ceramic material containing at least 90 w.% of oxynitride of silicium and aluminium in a single phase, displaying a crystalline structure like that of beta-silicium nitride, but with increased cellular dimension and in accordance with the formula claimed in Patent (170/11). The final ceramic material is developed from an intermediary ceramic material, which contains the oxynitride of silicium and aluminium in proportions differing from the given formula in that silicium and aluminium oxynitrides are enriched with Al and N as compared to the given formula.

Patent (170/15) of Lucas Industries refers to the production of a ceramic material by forming a powdery mixture, containing a first component (silicium, aluminium, oxygen, nitrogen) which reacts in such a manner in the last stage of sintering that a ceramic material is formed according to the formula:

$$Si_{6-z}Al_zO_2N_{8-z}$$

wherin $z = 0.38-1.5$, the mixture also containing a second component in the form of an oxide of yttrium, scandium, cerium, lanthanum or a member of the lanthanide group. This mixture is sintered under a protecting atmosphere (with or without pressure) at 1600°-2000^oC during a period of time lasting from 10 minutes to 5 hours.

According to Patent of Lucas Industries (170/16) elements of cutting tools can be produced by forming for example an edge of a ceramic material, containing min. 75 vol.% of a component (in one phase) having a beta-silicium nitride reticulated structure and corresponding to the known formula: $Si_{6-z}Al_zN_{8-z}O_2$, (wherein $0 < z < 5$), in combination with a second phase, containing a metal, for example yttrium, lanthanum, cerium, scandium or another earth-alkali metal.

The National Institute for Researches in Inorganic Materials (JP) disclose in Patent (188) a process for producing sialon sinter products according to the formula:

$$Si_{6-z}Al_zO_zN_{8-z}$$

wherein z = 0 to 4.2. The body made of the initial materials for sialon is sintered in a nitrogen gas atmosphere, while coating the surface of the body with a powder mixture containing 5-80 w.% SiO_2 and 20-95 w.% Si_3N_4 with sintering taking place at $1700°-1850°C$.

Patent (204/12) of Norton Company (US) aims at producing silicium oxynitride without applying a catalyst, based on a mixture of silicium and silicium oxide, in finely dispersed state, by heating in an inert (argon) atmosphere, to obtain an agglomerate (cake), without permitting the smelting of silicium for a period of time, necessary to form a silicium oxide layer on the silicium mixture particles, whereafter the mixture (to which nitrogen has been added) is heated to $1380°-1470°C$ to obtain the final product. The initial mixture may also contain less than 0.1 w.% of an alkali earth metal oxide.

Another Patent (204/16) of Norton Company refers to ceramic heating elements, composed of a sintered mixture, containing 10-60 w.% silicium carbide and 40-90 w.% of a material selected from silicium nitride, silicium oxynitride, silicium-aluminium oxynitride and/or their mixtures. The sintered mixture has a density of 90% of the nominal density value and a resistivity between 0.1 and 10^7 ohm.cm.

Patent (218/2) of The Research Institute for Special Inorganic Materials (JP) concerns the production of a heat-resistant ceramic body, produced by mixing 0.01 to 15 w.% of a polyborosiloxane, containing phenyl groups in side chains of silicon with a polysilane and polymerising the mixture to form polycarbosilane partly containing siloxane bonds, mixing the product e.g. up to 20% with a ceramic powder and sintering the mix after forming to a desired shape by heating at 800 to 2000°C in a non-oxidising atmosphere. The polysilane has the formula

$$\left[\begin{array}{c} R_1 \\ | \\ Si \\ | \\ R_2 \end{array} \right]_n$$

where n is at least 3, and R_1 and R_2 which may be the same or different
are methyl, ethyl or phenyl groups or hydrogen atom. The ceramic powder
is an oxide, carbide, boride, nitride or silicide. The ceramic body
may be used as building material, as aircraft component, motor car part
and for the manufacture of machine parts.

Süddeutsche Kalkstickstoff-Werke A.G. (DE) provide in Patent (243/1) a
process for producing refractory structural elements, consisting of
85-98 w.% silicium carbonitride, siliconitrides of calcium, magnesium,
strontium, barium, aluminium and/or iron, 1-10 w.% bonded carbon,
1-15 w.% silicium nitride and technological impurities. The major metal
content consists of calcium. The composition, which may also contain
reaction accelerators (oxides and/or fluorides of calcium, magnesium, iron
and/or manganese) is heated to 800°-1500°C. The nitriding gas consists
of nitrogen and/or ammonia and an inert component.

The object of Patent (256/2) of Toshiba Ceramics Ltd. (JP) is a process
for producing sintered bodies on the basis of β'-sialon, whereby a
pulverulent initial mixture is prepared from 10-1000 w.p. metallic Si-
powder, 20-80 w.% of silicium dioxide powder and 80-20 w.% aluminium
powder, this mixture being thoroughly agitated and formed into a green
body, which is sintered in nitrogeneous non-oxidising atmosphere at
1200°-1550°C.

10.3.3 Silanes based on silicon oxynitrides

Nippon Steel Corporation (JP) disclose in Patent (199/2) a chute for
blast furnaces, which comprises, at least in a layer on its inner surface,
carbon bricks, displaying pores of reduced dimensions, caused by forming
therein an Si-N-O composition. The pore size is under 5 microns. The
thickness of the inner surface layer corresponds to 1/3 - 1/2 of the total
wall thickness of the chute.

Patent (204/18) of Norton Company describes a porous refractory body,
the porosity of which is represented by the ratio between the output
"F" and the loss of charge "P" according to F/P higher than 26, in case
F is measured in $dm^3/hr/cm^2$, while P is measured in cm of water per cm
thickness, the pores of this body containing no fibres of Si_3N_4. The

major part of Si_2ON_2, forming the porous body, is present in form of
grains of coarse granulometry. The grains are agglomerated by more
dense Si_2ON_2, which covers the grains in a thickness of 20-100/u. The
dense coating is formed by nitriding (in situ) colloidal silicium oxide
and silicium.

In Patent (204/20) Norton Company reveal another porous (permeable)
refractory element, composed of refractory grains selected from silicium
oxynitride, mullite, zirconium, silicium oxide smelt, silicium nitride
or their mixtures, the grains being bonded by a bonding agent composed
of a mixture of silicium, a source of silicium oxide, calcium fluoride,
magnesium fluoride, barium oxide, calcium oxide, magnesium oxide, strontium oxide, cerium oxide, yttrium oxide or their mixture, the bonding
agent being applied in a proportion of 0-3% of the refractory component's
weight.

10.3.4   Miscellaneous

Kali Chemie A.G.(DE) claim in Patent (146) a process for producing silicic acid gel, with a very small amount of impurifications (like iron,
aluminium, arsenic and load). The gel is obtained by treating the gel
particles (obtained by a conventional process) with hydrochloric acid,
which may contain a substance for bonding the impurities. After this
treatment the product is washed in water and, as the case may be, dissolved in solutions of alkaline reaction.

Patent (167) of Laporte Industries Ltd. (GB)discloses   aqueous dispersions, containing a source of cation and having dispersed therein a
synthetic material having a structure characteristic of a smectite clay
mineral. The dispersions show advantageous properties if the synthetic
material contains iron ions bound in the structure thereof. The synthetic material may have a hectorite structure; it may contain up to
1.5 atoms of Fe per unit cell, and be manufactured by known processes,
suitably modified by the inclusion of a soluble iron salt. The dispersions may have a pH below 4 or may have a high content of metal or ammonium cations, and be used in gelled acidic or electrolyte containing
systems.

Lucas Industries Ltd. developed a method (170/17) producing a dense
ceramic product by mixing a ceramic powder containing a major proportion of an alpha-phase substituted silicon nitride with the formula:

$$M_x(Si,Al)_{12}(O,N)_{16},$$

where x is not greater than 2 and M is a modifying cation (e.g. yttrium,
calcium, lithium, magnesium, cerium) with at least one nitride of silicon and/or alumina, wherein the alumina is not more than 10 w.% of the
mixture, and sintering the mixture in a non-oxidising atmosphere at a
temperature of $1700^{\circ}C$ to $1900^{\circ}C$ so as to produce (1) a dense ceramic
product consisting essentially of beta-phase substituted silicon nitride
having the general formula:

$$Si_{6-z}Al_zN_{8-z}O_z,$$

where z is greater than zero and not greater than 1.5, and 0.05% wt to
20% wt of a further phase containing said modifying cation M, or (2) a
dense ceramic product consisting essentially of said beta-phase substituted silicon nitride , 0.05% wt to 90% wt of an alpha-phase substituted
silicon nitride obeying said formula $M_x(Si,Al)_{12}(O,N)_{16}$ and a minor
amount of a further phase containing said modifying cation M. The ceramic powder is preferably produced by effecting a first stage nitriding
by heating a powder mixture comprising silicon, aluminium, alumina and an
oxide or nitride of a modifying cation in a nitriding atmosphere at a
temperature below the melting point of aluminium until said first stage
nitriding is substantially complete; effecting a second stage nitriding
by heating the first stage nitrided mixture at a temperature below the
melting point of silicon, whilst maintaining a nitriding atmosphere until the second stage nitriding is substantially complete whereby a
friable material is produced, and subsequently sintering the comminuted
friable product at a temperature between $1650^{\circ}C$ and $1900^{\circ}C$, whilst maintaining a non-oxidising atmosphere to obtain a friable ceramic material
including more than 50 w.% of an alpha-phase substituted silicon nitride,
according to the formula $M_x(Si,Al)_{12}(O,N)_{16}$, where x is not greater than
1 and M is the modifying cation, or a friable ceramic material including
greater than 50 w.% of a mixture of said alpha-phase substituted silicon
nitride and a beta-phase substituted silicon nitride of the formula
$Si_{6-z}Al_zN_{8-z}O_z$, where z is greater than zero and not greater than 1.5.

Patent (171) of <u>The Macaulay Institute for Soil Research</u> and, essentially the Patent (189/3) of the <u>National Development Corporation (GB)</u>
concern an inorganic material, akin to imogolite, having a fibrous
tubular structure and obtained by preparing an aqueous hydroxyaluminium
silicate solution, containing up to 0.5 molar aluminium and of pH 3.1
to 5.0 by acid-hydrolysing aluminium alkoxides and (introduced not ear-
lier than the alkoxide) tetralkyl silicate. The solution thus prepared
is digested, e.g. at 40 to 170°C, until a product is obtained displaying
discernible electron diffraction peaks at 1.4 Å, 2.1 Å and 4.2 Å. The
product may be used as a support medium in electrophoresis, a molecular
sieve, a catalyst support, a catalyst, a coagulant, a sorbent, a thickener,
a hydrophiliser or a coherent film.

The object of Patent (204/19) of <u>Norton Company</u> is a composite, rotary
ceramic element, displaying an inner section and an outer section, the
inner section having a higher density and being made of silicium nitride,
while the outer section is composed of a mixture of silicium nitride and
a material, selected from silicium carbide, mullite, zirconium, yttrium
oxide, thorium oxide, tantalum carbide, alumina, titanium carbide and
their mixtures. The outer section of the element is provided with a corrosion and erosion resistant surface layer.

According to Patent (228/4) of <u>Shell Internationale Research Maatschappij</u>
crystalline silicates are produced having (a) a certain specific X-ray
powder diffraction pattern; (b) a composition expressed in moles of the
oxides $p(0.9 \pm 0.3)M_{2/n}O \cdot p(aX_2O_3 \cdot bY_2O_3) \cdot SiO_2$, wherein M = H and alkali
metal and/or alkaline earth metal, X = rhodium, chromium and/or scandium,
Y = aluminium, iron and/or gallium, a ⊃ 0.5, b ⩾ 0, a+B = 1, 0<p<0.1
and n = the valency of M. The crystalline silicates can be used, for
example, as catalysts in chemical conversions.

Patent (267/22) granted to <u>UKAEA</u> refers to a porous silicium nitride
body, the pores of which are filled partly or completely with an adhesive,
refractory filling material for reducing the gas permeability of the
body. The filling material consists of cordierite and boron silicate
glass, or of cordierite in combination with silicium oxynitride. Cordierite is present in an amount of 10-15 w.% (with regard to the silicium
nitride weight). The filling material contains boron silicate glass,
which comprises 5-20 w.% SiO_2 with regard to the silicium nitride weight.

CHAPTER 11

Intermediates mainly based on zirconium or zirconium-
containing compositions

Asaki Glass Co.Ltd. (JP) developed in Patent (19/1) a refractory
material, containing 40-78 w.% of ZrO_2 + SnO_2; 10-58 w.% of Al_2O_3 +
Cr_2O_3 and 2-20 w.% of SiO_2, the ZrO_2 content being higher than that of
SnO_2, the content of the latter surpassing that of SiO_2. More particularly the product consists of baddeleyite, corundum and a vitrified
phase, which fills up the interstices between the crystals of baddeleyite and corundum. The refractory material displays high corrosion
resistance.

J.H.F. Blake (GB) claims in Patent (33) a method of producing finely
divided metal oxygen-containing compounds (powders) comprising: a)
forming an intimate mixture in solution of a carbohydrate material and
at least one compound of e.g. zirconium; b) drying the solution of
step a) and igniting the resultant carbohydrate material/metal compound
mixture to produce fragile agglomerates of metal oxygen-containing compound particles. c) comminuting the agglomerates of step b) to produce
finely divided metal oxygen-containing compounds. The carbohydrate material may be sugar (sucrose), invert syrup, while the Zr solution may
also contain a compound of a metal that forms an oxide which stabilises
Zr, chosen from the Group IIIB metals, the Group IVB metals, niobium,
tantalium, the GroupVIB metals, manganese, iron, cobalt, nickel, copper,
zinc, cadmium, aluminium, gallium, tin, lead and bismuth. The fine
powder obtained by this process can be pressed into a shaped body, which
then is sintered at a temperature above $1000^{\circ}C$.

Patent (39/2) granted to Robert Bosch GmbH concerns the production of
stabilised zirconium oxide ceramic materials, containing silicate fluxing

agents and as stabilising substances calcium oxide and magnesium oxide
in an amount of 15-20 mol.% and at a mol. ratio CaO:Mg of 6:1-0.75:1.
Up to 5 mol.% of calcium oxide can be substituted with yttrium oxide
and/or ytterbium oxide. The silicate fluxing agent is applied in an
amount of 0.5-5 w.% (2-4 w.%) referring to the sum of the rest of oxides
applied.

Another Patent (39/3) of Robert Bosch GmbH refers to ceramic zirconium
oxide material, similar to that of the former patent and suitable for
use as a solid electrolyte and which consists of ZrO_2, a stabilising
oxide like yttrium oxide (U_2O_3) or ytterbium oxide (Yb_2O_3) or CaO or
MgO (alone or in mixtures), furthermore Al_2O_3 in an amount of 8-25% of
the volume of all other components. Prior to sintering, Al_2O_3 displays
a specific surface of $> 1 \ m^2/g$.

The Chas.Taylor's Sons Co. (US) reveal in Patent (58/1) a refractory
mass,containing 70.0-96.8 w.% zirconium, 3.0-29.8 w.% iron-chromite ore
and 0.2-5.0 w.% rutile-titanium dioxide. The mass is prepared by mixing
together the aforementioned components, the mixture being formed into a
ceramic body and fired at a temperature between 1450 and $1650^{o}C$. The
obtained mass displays a high rupture modulus and a low porosity.

Another Patent (58/3) of Chas. Taylor's Sons,Co. concerns a process
for preparing refractory articles (for example tapping tubes for metals)
wherein a mixture is prepared of a refractory composition and an aqueous
suspension of colloidal silica for obtaining a pulp, which then is fed
into a mould and heated in order to solidify the pulp, the solidified
article then being removed from the mould, dried and calcined. More
particularly, the refractory composition contains: 45-75 w.% of a fusion
product, 10-30 w.% of disthene, 3-15 w.% of zirconium, 3-12 w.% of calcined alumina, 0.5-10% of chrome oxide, while the fusion product is a
mixture of alumina and zirconium.

Patent (65/1) of Commonwealth Scientific and Industrial Research Orga-
nisation (Australia) refers to zirconium oxide ceramic material, which
is partly stabilised with 3.3-4.7 w.% calcium oxide. The body displays
a double-phase flaky microstructure, essentially consisting of metastable, tetragonal phases of cubic matrix particles, which are converted

irreversibly into normal monocline form, when the body (the article made
of such body) is exposed to very high loads.

In another Patent (65/3) the <u>Commonwealth Scientific and Industrial
Research Organisation</u> describes a magnesia partially stabilised zirconia ceramic material, possessing from about 2.8 to about 4.0 w.%
magnesia, and made from a zirconia powder containing not more than about
0.03% silica. The ceramic material has a microstructure, produced as
a consequence of the method by which the material is made, which provides both high strength and good thermal shock resistance properties.
This microstructure comprises grains of cubic stabilised zirconia,
within which are formed, during cooling from the firing temperature,
precipitates of tetragonal zirconia. The precipitates are elliptical
in shape, with a long axis of about 1500 $\overset{o}{A}$ units. Additionally, some
of the tetragonal zirconia precipitates are made to transform into a
non-twinned microcrystalline monoclinic form of zirconia by reducing
the temperature of the material to below 800°C, then subsequently holding the material at a temperature in the range from 1000°C to about
1400°C. The ceramic material may also contain up to 36 w.% of hafnia.
Such materials can be used for moulds, dies, tappet facings and cutting
tools.

Patent (70/4) of <u>Corning Glass Works (US)</u> reveals a process for producing
a sintered ceramic material by composing a charge, containing zirconium
by shaping the charge into a green body and sintering it. The charge
should contain zirconium hydroxide, amounting to 70 w.% of the ceramic
charge. Sintering takes place at a temperature over 1400°C, at which
zirconium in the ceramic charge is completely stabilised. After extrusion
a product of alveolar, monolithic structure is obtained.

Patent (82/1) of <u>E.I. Du Pont de Nemours and Company (US)</u>relates to particulate abrasive compositions containing 69-94 w.% of a matrix, composed of zirconium carbide and titanium carbide, titanium boride and zirconium carbide with 6-31 w.% of crystalline titanium diboride particles,
dispersed in the matrix. Titanium diboride is present in the form of
particles of 0.5-30 microns, while the particle size of the abrasive
material is between 37 and 840 microns, the apparent density of abrasive
particles being between 4.8 and 5.3 g/cm^3.

The object of Patent (84/2) of Dynamit Nobel A.G. (DE) is a refractory
dry stamping mass, on the basis of partly stabilised zirconium oxide
for the lining of furnaces, used in smelting high-alloy noble steels,
nickel-base alloys, chrome-nickel base alloys and special alloys with
improved resistance against temperature changes at about $1650^{\circ}C$. The
lining also displays an increased resistance against sintering by the
feed-in charge and high temperatures.

In a process, revealed in Patent (84/3) of Dynamit Nobel A.G. for producing basic zirconium carbonates, composed of pure zirconium salts, a
zirconium sulphate $Zr(SO_4)_2$. $4H_2O$ or zirconium oxy-chloride $ZrOCl_2$.
$8H_2O$ are converted with an ammoniacal solution, the forming basic salts
then being separated and transferred into basic zirconium carbonate,
which can be calcined into an acid-soluble and finely dispersed powder.

Foseco Trading Co. (CH) disclose in Patent (100/2) a refractory support
composition, formed on at least one of the supporting faces of articles
or coating, composed of zirconium, fragmented by plasma jets into a
granulometry below 0.053 mm. The coating layer on the support surface
displays a thickness of 0.1-1.0 mm. The composition also contains a
binder, composed of a natural or synthetic resin, cellulose ether, poly-
acrylamide, polyvinylalcohol or polyethylene glycol.

Patent (110/1) of The Goldschmidt A.G. (DE) relates to a process for
preparing calcium zirconate with a content of free calcium oxide of
< 0.2 w.%, through the calcination of zirconium oxide and calcium carbonate, suitable for use in the production of shaped bodies (stones,
nozzles, slide valves) of steel by continuous casting.

Patent (110/2) of Th.Goldschmidt A.G. concerns the production of refractory materials on the basis of stabilised cubic zirconium oxide through
sintering with calcium zirconate.

In Patent (124/3) Hitachi Ltd. disclose ceramic materials for an oxygen
sensing element of solid electrolytes of a zirconium oxide-yttrium
oxide system, the ceramic material containing aggregates of cubic zirconium oxide particles (average particle size: 2-10 μm) and aggregates
of monocline zirconium oxide particles (average particle size: 0.2-1 μm)

the aggregates of cubic zirconium oxide particles being in mutual contact, while the monocline zirconium oxide particles are distributed, as
aggregates, in interstices between the aggregates of zirconium oxide
particles.

L'Institut National Interuniversitaire des Silicates, Sols et Matériaux
(I.N.I.S.Ma) and Service de Science des Matériaux de l'Université de Mons
(Be) reveal in Patent (133) a production process of refractory products
on the basis of zirconium, tabular aluminium and additives, like a
fluxing agent and/or a binder and/or a lubricating substance,for example,
soda-containing carboxy-methylcellulose, soda-silicate, substituted cel-
luloses, alkali-organic compositions. The ceramic mass is sintered during 15-66 hours, according to a programmed schedule in the $1540^{\circ}-1760^{\circ}C$
temperature range.

Patent (138) of The Institut Vysokikh Temperatur Akadamii Nauk SSR (USSR)
relates to a heat-resistant porous material, comprising 50-75 vol.% of
microspheres of a high-melting point oxide,like zirconium dioxide, having
a melting point in excess of $1700^{\circ}C$, the microspheres being 10-200 m/u
in diameter and being sintered directly to each other so that the diameter of the circle of contact between the microspheres is 0.2-0.5 of
the microsphere diameter. The heat-resistant porous material may additionally incorporate a filler which partly fills the pores of the material, and which may be a metal, an alloy, or an intermetallic compound
or a phenol-formaldehyde resin, a polyvinyl alcohol, an epoxy resin, or
an organosilicone polymer or glass. Due to the strength of the porous
material in accordance with the invention, it is possible to obtain
various kinds of special purpose materials for diversified applications,
such as in high-strength heat-resistant electrical engineering, and
heat-insulating and thermostable materials, ensuring stable operation
of various structures without changes in geometry.

Keeling & Walker Ltd. (GB)propose in Patent (148) a process for producing
a refractory body, according to which plasma-dissociated zirconium sand
is processed into powder, which then is shaped and exposed to sintering,
thereby obtaining a homogeneous body. Prior to the shaping phase, a
functional additive is incorporated into the powder, the additive containing iron oxide, iron hydroxide, lime, magnesium, yttrium, aluminium

oxide, mullite. The composition may also contain agglomerating agents like water-soluble polymers, clays, fluxing substances. The sand particles are of a size, enabling them to pass through a sieve, having 79-138 meshes per cm.

Patent (154/1) of Kombinat VEB Keramische Werke Hermsdorf (GDR) concerns the production of sintered, non-porous bodies, of high mechanical resistance and suitable to forming glazings, from zirconium and glass, at temperatures under 1050°C. More particularly the bodies consist of: 55-65 mol.% SiO_2; 55-65 mol.% Al_2O_3; 12-18 mol.% MgO; 7-20 mol.% CaO; 4-10 mol.% B_2O_3; 0-10 mol.% BaO; 0-3mol.% ZnO; 0-1 mol.% R_2O and 0.5-0.3 mol.% of a refining agent. The body displays an elasticity modulus higher than 0.8 . $10^6 kp/cm^2$ and a linear thermal expansion coefficient under 4.5 . 10^{-6} degree^{-1} in the temperature range between 20 and 400°C.

Patent (160) of KSR International Ltd. (GB) provides a ZrO_2/Mg_2SiO_4 refractory formulation suitable for firing to produce ceramically-bonded bodies capable of resisting high temperatures, molten metal and thermal shock, or to form the basis of low-temperature curing concretes having these characteristics, consisting of a blend of two granulated, prefired refractory compositions, each consisting predominantly of ZrO_2 and Mg_2SiO_4. The ZrO_2 in one composition is in the shock-susceptible unstabilised monoclinic form while it is in the MgO-stabilised cubic form in the other composition. The blend can be on a 1:1 weight basis, although usually the composition containing the stabilised ZrO_2 will predominate. Each composition is conveniently prepared by firing together $ZrSiO_4$ and MgO at 1550°C or higher; for the first shock-susceptible composition the MgO will be present in a stoichiometric or substoichiometric amount and for the second, shock-resisting composition the MgO will be present in a super-stoichiometric amount.

The Max-Planck-Institut für Eisenforschung GmbH   propose in Patent (176/1) a solid electrolyte with ion-conducting capacity at high temperature, on the basis of zirconium oxide and calcium oxide. More particularly the electrolyte consists of

$$CaO_{0.85 \text{ to } 1.10} \cdot ZrO_2$$

The CaO component can be substituted completely or partly by MgO and/or Y_2O_3. The solid electrolyte can be used in galvanic cells for determining the partial oxygen pressure of gases and the oxygen activity of metal smelts.

In Patent (176/9) the Max-Planck Gesellschaft reveal a ceramic body, composed of a compact ceramic matrix and produced by sintering a powdery mixture, which contains the material of the matrix and a component, comprising ZrO_2 and/or HfO, followed by a heat treatment at a temperature above the sintering temperature. The matrix may be composed of mullite, forsterite or cordierite. The ceramic body displays high resistance against thermal shocks and excellent mechanical strength.

Patent (186/1) of Murata Manufacturing Co.Ltd. (JP) concerns dielectric masses for high frequencies, consisting of an initial mass, containing 22-43 w.% titanium dioxide, 38-58 w.% zirconium oxide, 9-26 w.% tin (IV)-oxide, furthermore an additive, composed of lanthanum oxide or a blend of lanthanum oxide and zinc oxide, the additive forming 0.5-10 w.% of the total amount of the initial mass and additive.

Another Patent (186/2) of Murata Manufacturing Co.Ltd. also provides dielectric ceramic masses for high frequencies, consisting of an initial mass (90-99.8 w.%), which, in addition to the components indicated in (186/1) contains, as additive, 0.2-10 w.% cobalt (III)-oxide, and max. 7.0 w.% zinc oxide.

According to Patents (200) Nippon Telegraph & Telephone Public Corp. and Nippon Tsushin Kogyo K.K. (JP) ceramic materials suitable for use in semi-conductors are produced by mixing $ZrSiO_4$ with a calcination product of clay, like kaolin, figuline and $(Al_2O_3 . 2 SiO_2)$, the mixture also being added an additive selected from the group that contains LiO_2, TiO_2, ZnO or the group that contains CoO, Co_2O_3, Co_3O_4, MgO, this mixture being formed and the shaped article obtained being sintered at 1200°-1450^oC.

Patent (204/13) of Norton Company refers to the production of abrasive grains on the basis of zirconium oxide, the composition being free of pores and containing 10-70 w.% ZrO_2; 5-25 w.% Cr_2O_3; 0-85 w.% Al_2O_3, furthermore up to 5% impurities (oxides), which may only contain a very small amount of soda or metallic chrome in free state.

In Patent (209/3) the Plessey Co.Ltd. (GB) provides a ceramic material which comprises lead zirconate substituted with uranium and wherein zirconium is also substantially equally substituted with either iron and niobium, iron and tantalum, or nickel and niobium. When zirconium is substituted with either iron and niobium, iron and tantalum, or nickel and niobium in equal amounts,the following formulae apply:

$$Pb \left\{ Zr_{(1-2x)} Fe_x Nb_x \right\}_{1-w} UO_{w3}$$

$$Pb \left\{ Zr_{(1-2x)} Fe_x Ta_x \right\}_{1-w} UO_{w3}$$

$$Pb \left\{ Zr_{(1-2x)} Ni_x Nb_x \right\}_{1-w} UO_{w3}$$

The ceramic composition may include bismuth and potassium, substituted for the lead, according to the formula:

$$Pb_{1-\frac{3y}{2}-\frac{z}{2}} Bi_y K_z$$

The ceramic materials according to this patent have pyroelectric properties and are, therefore, ideally suited for use in infra-red devices, such as infra-red detectors and detector arrays, or in thermal imaging systems such as infra-red television cameras.

Société Européenne des Produits Réfractaires (FR) present in Patent (237/3) a process of forming balls (pellets) of a ceramic material,by fusion and solidification of an initial charge, containing ZrO_2, SiO_2 with a Zr_2/SiO_2 ratio higher than 1.5; furthermore also containing: Al_2O_3 with an Al_2SO_3/SiO_2 ratio between 0 and 1.5; Na_2O with an Na_2O/SiO_2 ratio between 0 and 0.04; MgO with an MgO/SiO_2 ratio between 0 and 1 and finally CaO with a CaO/SiO_2 ratio between 0 and 1.45.

Another Patent (237/4) of Société Européenne des Produits Réfractaires concerns an amorphous composition, comprising a vitrified phase, and as main components: ZrO_2; Al_2O_3; SiO_2, further as the case may be: Cr_2O_3 and 0-5% of a hydraulic cement. The refractory composition (after

smelting) contains 75-85% ZrO_2; 2-8% SiO_2; 9-17% Al_2O_3; 0.5-0.7% other
oxides (and their mixtures), finally a surface active agent. The refractory material is suitable for use in the base of glass-smelting
furnaces.

Yet another Patent (237/6) of Société Européenne des Produits Réfractaires
refers to petroleum extraction processes. More particularly, it provides
a granulated supporting material of high mechanical stability of a volume-
weight less than 3250 kg/m^3, from which balls (pellets) are made by fusion,
granulation and solidification of an initial composition, which contains
46-50 w.p. ZrO; 54-50 w.% SiO_2 (the sum of ZrO_2 + SiO_2 amounting to 100
parts), further 0-19 w.% Al_2O_3 and 0-13.5 w.% of an oxide of Mg, Ca, Fe
and Ti.

Patent (245/2) of Sumitomo Aluminium Smelting Co.Ltd. (JP) discloses a
method of producing a sintered zirconia article, which comprises: mixing
(a) 10-50 w.% of partially stabilised zirconia powder, obtained by sintering an admixture of zirconia powder and stabilising agent therefor and
pulverising the sintered mixture, and (b) 90-50 w.% of zirconia powder
and stabilising agent therefor, shaping the resulting mixture, sintering
the shaped mixture, and then heat-treating the sintered mixture at a
temperature from 1200° to 1500°C to obtain a sintered zirconia article
not more than 70 w.% of which is of cubic phase crystal structure and
which is constituted by 5-40 w.% of particles of 1 to 30/u diameter.
The sintered article has sufficiently high density and thermal shock
resistance for use as an oxygen sensor.

In Patent (254/5) of Tokyo Shibaura Denki K.K. the production is claimed
of an engine part, comprising at least partially a sintered body which
mainly consists of zirconium and yttrium, the yttrium being contained
in the sintered body in an amount of 0.1 to 15 mol.% with respect to the
total of zirconium and yttrium. The sintered body has a surface rough-
ness of not greater than 5S, a fracture toughness K_{IC} of 6 to 15 MPa.m$^{1/2}$,
a porosity of not greater than 10%, a grain size of not greater than
5/um, a coefficient of thermal expansion of 6 to 13 x 10^{-6}, a thermal
conductivity of 1 to 5 W/m .°C; a flexural strength of 200 to 1,500
MPa; a thermal shock resistance (with a critical thermal shock temperature difference of 300 to 500°C); a Young's modulus of 150 to 250 GPa
and a density of 5.3 to 6.2 x 10^3 kg/m^3.

The object of Patent (266) of <u>United Aircraft Corporation (US)</u> is a composite article of refractory oxide, obtained by controlled solidification. The composite article comprises a matrix phase, consisting of calcium oxide and zirconium oxide, furthermore a reinforcing phase of zirconium oxide, this phase being embedded in the matrix in the form of lamellas, oriented in the direction of the main tension, which is expected to develop in the article.

A colloidal dispersion is described in Patent (267/15) of <u>UKAEA</u>, suitable for use in the preparation of a material containing zirconium, silicon and oxygen, by heating a gel precursor for the material to convert it to the material, the precursor being a gel-containing silica and zirconium or a precursor for zirconium. The gel precursor may be prepared by gelling a colloidal dispersion. Pigments may be formed by inclusion of pigmentation agents e.g. Pr(III), fluxes such as NaF or CaCl, metal powder and/or metal oxides, e.g. TiO_2. The material may contain Zr, Si and O in a 1:1:4 atom ratio.

<u>Th. Goldschmidt A.G.</u>   disclose in Patent (110/3) a process for producing zirconium oxide, suitable for use in electro-ceramics, and using as initial substance calcium zirconate, which is dissolved in hydrochloric acid and to which fluorine is added in an amount of 0.1-0.1 mole fluoride per 1 mole of the zirconium component (using for this purpose an alkaline or earth-alkaline element). Then a solution of sulphuric acid is added to the mixture,which is heated to $80^{\circ}C$ (or higher), the mixture being held at this temperature for 10 minutes. The forming suspension is left resting for 2 hours, followed by filtration and the washing of the sediment, into which a solution of ammonium carbonate is introduced, in equimolar amount with regard to the zirconium composition. Thereafter carbonaceous gas is added to the solution, the sediment is filtered, washed, dried and sintered at $1000-1100^{\circ}C$.

CHAPTER 12

Intermediates mainly containing boron

12.1 Boron nitrides and carbides

Patent (25) of the Battelle Development Corporation (US) provides a
method for the deposition of cubic boron nitride on a substrate, by
activated reactive evaporation. The method includes: supporting and
heating a substrate in vacuum; evaporating metal vapours into a zone
between the substrate and the metal alloy source, from a metal alloy
source consisting essentially of at least 60 w.% to balance of boron,
with from 2 to 12 w.% of aluminium and at least 0.2 to 24 w.% of at
least one of cobalt, nickel, manganese, or other aluminide forming
element; introducing ammonia gas into the zone and generating an electrical field in the zone for ionising the metal vapours and gas atoms
in the zone. The substrate is generally heated to a temperature of at
least $300^{\circ}C$ (with preferred substrate temperatures between about $500^{\circ}C$
and $1100^{\circ}C$), the ammonia gas pressure preferably is about 1×10^{-4} torr
to 8×10^{-3} torr, and plasma activation in the zone desirably may be
effected by employing deflection electrode maintained at a positive
voltage potential and positioned between the substrate and the source
of evaporating metal vapours.

According to Patent (28/2) of De Beers Industrial Diamond Division
(South Africa) an abrasive agglomerate can be obtained, containing cubic
boron nitride, diamond and their mixture, bonded together by a matrix of
refractory material and a solvent, capable of dissolving, at least partly, the abrasive particles. The refractory material consists of a boride,
a nitride or a silicate, while the solvent applied is a metal (e.g. Al)
or metal alloy. The weight ratio between the solvent and the refractory
material is established between 0.33:0.67 and 0.67:0.33.

In Patent (51/4) The Carborundum Company reveal a liquid sintering aid
for densifying ceramic material, selected from solutions of H_3BO_3, B_2O_3
and mixtures of these solutions in sintering ceramic articles, e.g.
silicon carbide, a shaped green body is formed from a particulate ceramic material and a resin binder, and the green body is baked at a temperature of 500 to 1000°C to form a porous body. The liquid sintering
aid of B_2O_3 and/or H_3BO_3 is then dispersed through the porous body and
the thus treated body is sintered at a temperature of 1900° to 2200°C,
to produce the sintered ceramic article.

Another Patent (51/16) of The Carborundum Company refers to the production of boron nitride articles by mixing 2-40 w.% boron oxide with
60-98 w.% fibres of boron nitride or fibres of boron oxide, which are
partly nitrided; by forming from these materials a shaped body,
heating it in an anhydrous gas (ammonium nitrogen) atmosphere at a
temperature above the fusion point of boron oxide, but below the fusion
(decomposition) point of fibres. Heating should last until at least
part of boron oxide is stabilised in the pores. The fourth production
phase consists of heating at a temperature ($460-1400^{\circ}$C) for transforming boron oxide into boron nitride.

Patent (74/3) of Denki Kagaku Kogyo K.K. concerns the production of corrosion resistant high-temperature shaped bodies, obtained by hot pressing
a mixture of 1-10 w.parts of aluminium powder and 100 w.parts of a powdery mixture of an electrically conductive, refractory material of boron
nitride and aluminium nitride. More particularly the mixture contains
40-70 w.% boron nitride, 30-60 w.% of aluminium nitride. The main components of the electrically conductive material are TiB_2; ZrB_2; TiC;
ZrC or a mixture thereof.

Abrasive particles are produced according to Patent (82/2) of E.I.Du Pont
de Nemours and Company which are composed of a matrix of boron carbide
and boron crystals, this composite system also containing a solid solution, which comprises titanium carbide (40-75 mol.%), zirconium carbide
(20-35 mol.%), tantalum carbide (5-25 mol.%) and crystalline titanium
diboride (2-35 mol.%).

Patent (104/19) of General Electric Company (US) refers to an element

of a tool, composed of bonded particles of diamond and cubic boron nitride
representing 70-95% of the element's volume, this composition also containing a network, which interconnects the pores dispersed in the element,
these pores forming 5-30% of the element's volume and about 0.05-3% of
the volume of a metal phase, consisting of an agent promoting the sintering of the diamond and boron particles. The metal phase may contain
various combinations of catalytic (Cr, Mn, Ta) and non-catalytic metals
and their alloys.

Patent (104/20) of General Electric Company discloses the production
of cubic polycrystalline boron nitride bodies by: placing hexagonal
pyrolytic boron nitride in the form of a disc, (with chamfered edges)
in a reaction vessel; compressing the vessel's content under a pressure
of 50-100 kbars; heating the vessel and its content to about 1800°C;
realising such pressure, temperature and holding time conditions,
which ensure the transformation of pyrolytic boron nitride into polycrystalline sintered boron nitride. The hexagonal pyrolytic boron nitride is obtained by "germination". The reaction vessel contains a
metal screen, surrounding the boron nitride during the transformation
phase and protecting it against contamination.

Patent (130) of Institut Fiziki Tverdogo Tela i Poluprovodnikov Akademii
Nauk Belorusskoi SSR (USSR) relates to an ultrahard material comprising
boron nitride and alloying substances composed of a solid solution of
an alloying substance, consisting of an element of group III of the Periodic Table, which is capable of forming tetrahedral bonds, the solid
solution containing the components in the following amounts (atom %):
boron 42-61; nitrogen 39-50; alloying substance 0.1-30; other additives 0.01-2; boron nitride being present in cubic form. The solid
solution of cubic boron nitride displays a crystal lattice parameter of
about 3,620 Å.

In Patent (131) the Institut Fiziki Vysokich Davleni; Akademii Nauk SSSR
(USSR) disclose a composite body, prepared on the basis of cubic boron
nitride, the crystals of which are bonded together with a bonding agent
consisting of intermetallic compounds, like Ti_2Cu, $TiCu$, Ti_2Cu_3, $TiCu_3$,
Zr_2Cu, $ZrCu$, Zr_2Cu_3, $ZrCu_3$, used individually or in combination, in an
amount of 10-35 vol.%, the rest consisting of cubic boron nitride.

More particularly, the composite body contains 80-90 vol.% cubic boron
nitride; 8-16 vol.% intermetallic compounds and 2-4 vol.% alloying additives.

Patent (134) of the Institut Novykh Khimicheskikh Problem Akademii
Nauk SSSR and Institut Khimicheskoi Fiziki Akademii Nauk SSSR (USSR)
describes a process for the production of polycrystalline boron nitride
by treating, under high pressure and at a high temperature, an initial
material, which contains boron nitride and high smelting point substances (for example monocrystalline powders of nitrides of transition metals) with a particle size between 50 and 1000 Å, in an amount of 0.1-30
w.%. As the refractory components TiC_xN_y (x = 0.1-0.9; y = 0.9-0.1)
or TiB_xN_y (x = 0.05 - 0.30; y = 0.95 - 0.70) can be applied. Prior
to being introduced in the charge, the boron nitride component may be
exposed to percussion effects.

The Institut Sverkhtverdykh Materialov Akademii Nauk Ukrainskoi SSR,
(USSR) developed a process (137/1) for producing ultrahard materials,
which contain diamond grains of less than 1 mm size by hot pressing the
mixture of diamond grains with a metal carbide, in a graphite mould at
$\pm$ 1800°C, the heating process being carried out in two phases: first
heating to 1200°C (at a rate of 1000-1100°C/minute) and then heating to
the final temperature (at a rate of 3000-6000°C/minute), the mould
being heated by high frequency induction heating. After having reached
the temperature of 1200°C, the charge is kept in the mould for not more
than 2 minutes.

Another Patent (137/2) of the aforementioned Soviet (Ukrainian) institute
reveals in Patent (137/2) an ultrahard cubic boron nitride, obtained by
sintering at 1600°C under a pressure of 50 kilobars. Each boron nitride
particle is completely coated with a crystalline substance according to
the formula:

$$B_x N_y C_z$$

(x, y, z = different values between 0 and 1), this crystalline substance
bonding the cubic boron nitride particles, which are exposed to a carbon
containing gas current (methane) for forming a carbon layer of 1Å-100Å
thickness on the surface of the boron particles, which then are sintered

at temperatures corresponding to the thermodynamic stability of cubic
boron nitride under a pressure of 50 kbars.

According to Patent (139/3) of I B M (US) a process of forming a refractory compound (boron nitride) consists of: evacuating a chamber,
containing the refractory compound structure, which is a combination of
a refractory or metallic element or both and a non-metallic element,
the refractory or metallic element being solid at room temperature and
having a vapour pressure at least one order of magnitude below that of
the non-metallic element, in the temperature range at which decomposition
of the refractory compound occurs; this structure having on its surface an adherent dense layer of the refractory or metallic element or
both of the refractory compound, and heating the structure in the chamber at a decomposition temperature for a time sufficient to form an
elemental layer of the refractory or metallic element or both on the
structure. The non-metallic element is evacuated from the chamber when
formed from the decomposition.

Kennecott Corporation (US) disclose in Patent (149/5) a process for producing non-woven, yet internally cohesive, boron nitride fibre batts,
suitable for use as an electric cell separator in a lithium-sulfide bat-
tery. Molten boron oxide is centrifugally spun into strands and atte-
nuated by relative humidity. The fibres are funnelled into a chamber
and subjected to a turbulent air flow, which causes random orientation
and intertwining of the fibres, resulting in multiple mechanical bonds.
The compacted boron oxide fibre bundle thus produced is heated in an
anhydrous ammonia atmosphere to convert boron oxide in the fibres into
boron nitride.

Patent (149/6) of the Kennecott Corporation also refers to non-woven
boron nitride bonded fibres for producing a mat, suitable for use as
electric cell separator in lithium-sulfide batteries. According to the
patent molten boron oxide is centrifugally spun into strands and atte-
nuated by gas stream into fibres which are compacted at a controlled
relative humidity into a batt and heated in an anhydrous ammonia atmosphere to convert boron oxide in the fibres to boron nitride. The
BN fibres are blended with a lesser amount of boron oxide fibres and
a non-aqueous liquid medium to form a slurry. The slurry is processed

through a Fourdrinier machine to form a felt, and the felt is calendered by passing it through the nip of a pair of calender rolls, at an appropriate temperature and pressure to soften the boron oxide binder to fuse the BN fibres together. The interstitial boron oxide then is converted to boron nitride. The process to obtain the BN fibres is represented in the following figure:

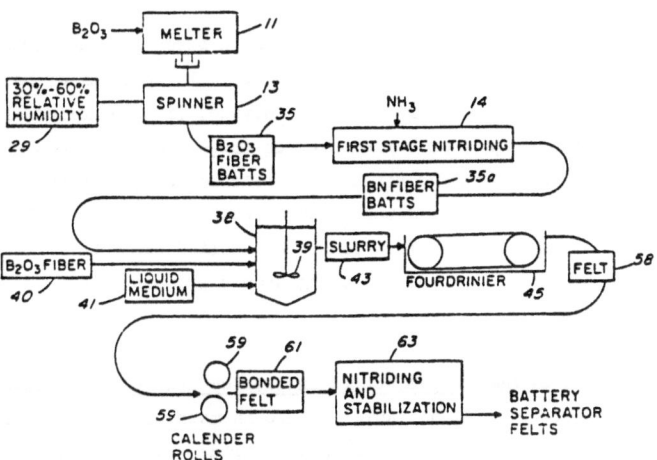

Lucas Industries Ltd. (GB) developed in Patent (170/3) a sintering process for the production of a silicon nitride article or a silicon nitride-containing article, wherein the material being sintered is protected with a medium, which contains boron nitride and which also provides a silicon monoxide vapour pressure greater than the silicon monoxide vapour pressure in the material being sintered at the elevated temperature of the process. The walls of the graphite component presented to the article are also provided with a protective coating to prevent reaction between the silicon monoxide and the walls of the component at the elevated temperature applied. The protective coating also includes alumina or aluminium nitride and a boron nitride spacer is interposed between the coating and the material being sintered.

According to Patent (227/2) granted to the Secretary of State for Defence (GB) boron nitride fibre material is produced by depositing a coating of boron nitride (BN) on carbon fibre and subsequently heating the coated fibre in a gas, e.g. air, which reacts with carbon to drive-off the car-

bon as a gaseous product and leave substantially hollow boron nitride
fibre, constituted by the coating. The BN may be deposited by reacting
boron trifluoride with ammonia at elevated temperature. The carbon
fibre may be a porous mass, such as a felt or fabric, and the resulting
BN fibre mass may be impregnated by further deposition of BN to produce
a dense, solid, BN-BN composite material.

Showa Denko K.K. (JP) claim in Patent (230) a boron nitride composition,
represented by the formula: $LiMBN_2$, (wherein L = calcium or barium)
obtained by forming a mixture of finely divided lithium nitride (LiN_3),
calcium nitride (Ca_3N_2) or barium nitride (Ba_3N_2) and hexagonal boron
nitride, in powder form and effecting a reaction between these components, in molten state, at a temperature between 800° and 1300^oC, followed by cooling and solidification. The applied elements: Li, Ca or
Ba; B and N are used in the following atomic proportions:
(1-1,4):(1-1,4):1:2.

Patent (246/3) of Sumitomo Electric Industries Ltd. (JP) provides a
sintered compacted body, suitable for use in machining tools, which contains 8-10 vol.% of high pressure boron nitride, the rest consisting of
a matrix, composed of carbides, nitrides, carbonitrides, borides and
silicides of transition metals (groups IVa, Va, VIa of the P.S.) and
their mixtures, in solid solution, the matrix forming a continuous bonding structure in the sintered body. The main component: boron nitride
is of the cubic type and is obtained by the transformation of the
wurtzite form of boron nitride during sintering.

Another Patent (246/4) of Sumitomo Electric Industries Ltd. describes
sintered blocks also suitable for use in machine tools and comprising
80-95 vol.% boron nitride, while the rest of the block mass consists of
a binder, e.g. a carbide, nitride or carbonitride of a transition metal
(IVa and Va groups of the P.S.), furthermore a mixture of the aforementioned compositions in solid solution. The binder may also contain
copper in combination with iron, nickel, cobalt, in an amount between
1-50 w.%.

Patent (279/2) of Vsesojuznii Nauchno-Issledovatelskii Institut Abrazivov
i Shlifovania (USSR) refers to the preparation of polycrystalline cubic

boron nitride by applying a pressure of 40-90 kilobars at a temperature of 1200°-2400°C to hexagonal boron nitride in the presence of a catalyst, for example a zinc composition, e.g. an oxide, hydroxide, nitride, amide of zinc or a mixture thereof, added in an amount 0.1-12 w.%.

12.2   Borides

Patent (27/2) of <u>Battelle Memorial Institute (CH)</u> reveals a powder composition, suitable for thermal sintering without pressure. The composition consists of titanium boride and is suitable for use in electrodes, which must be resistant to the corroding effect of aluminium smelts. The composition also contains a densifying additive of titanium hydride and boron hydride, the additive being applied at a ratio: 1 mole TiH_2 per 2 moles of boron.

According to Patent (113) of <u>Great Lakes Carbon Corporation (US)</u> TiB_2-carbon composites are produced by mixing the raw materials, composed of carbon, TiB_2, pitch, and other reactants, forming a shaped article, processing in a nitrogen atmosphere up to 2100°C, and in a noble gas above 2100°C by pressureless sintering of TiB_2 or other refractory hard metal powder, or by molding or extrusion of plastic mixes of binder and particulate carbon and refractory hard metal. The carbon powder is selected from the group consisting of graphite, calcined petroleum coke, metallurgical coke and charcoal. The article obtained after sintering is cooled, then impregnated under a pressure of 2 to 15×10^5 Pa at 175° to 250°C with a carbonisable impregnant, baked to carbonise the impregnant in a cycle rising to 700° to 1100° over a period of 1 to 10 days, then further heated to 1700° to 2400°C to form a TiB_2-graphite composite structure.

Patent (118) of <u>GTE Sylvania Inc. (US)</u> is concerned with electrically conductive boats of refractory composition for use in the vacuum evaporation of metals. In such boats titanium diboride is the electrically conductive material used, along with an insulating material (boron nitride), which imparts machinability and is resistant to molten metals such as aluminium. In such boats it is desirable that the curve of resistivity versus temperature be substantially flat over the operating temperature range of the boat, because such a flat curve aids in maintaining a balanced operation during a continuous evaporation operation, for example, where aluminium wire on a spool is unreeled and fed into a boat where it is evaporated and deposited on plastic film being unwound from one roll and taken up on another. The patent discloses an electrically conductive evaporating boat in which the curve of resistivity versus temperature is generally flatter than that of prior art boats. The boat composition consists essentially of a conductor, titanium, diboride and/or zirconium diboride, a coconductor, silicon carbide, with the balance being boron nitride. The composition may include a small amount of a flux, such as boron oxide. In order to attain the desired resistivity versus temperature curve, the composition should be about 15 to 35% by weight silicon carbide, 35 to 60 w.% titanium diboride and/or zirconium diboride, with boron nitride making up the balance.

PPG Industries, Inc.(US) developed a process (212/1) for producing articles of titanium diboride by cold-forming and sintering. The process includes disintegration of titanium diboride into particles of a specific surface under 1 m^2/g, in the presence of an adjuvant for crushing in liquid state (for example liquid hydrocarbon with 5-8 C atoms, or a chlorinated hydrocarbon with 2-4 C atoms). The articles display a high volumetric mass (95% of the nominal volumetric mass of titanium diboride).

Another Patent (212/2) of PPG Industries, Inc. refers to the production of articles by cold-forming and sintering a powder, composed of diborides of a group IVb metal, the powder containing 0.5-5 w.% of a hydrocarbon binder in dispersed state. The cold-formed article is placed in a casing, displaying thermal conductivity, from which impurities have been removed and which is chemically inert with regard to the article placed therein, whereafter the casing and its content are heated to the sintering temperature of the metal diboride, without applying any pressure. The thermally conductive casing can be made of graphite, of a diboride or a carbide of a group IVb metal. The metal diboride contains not more than 0.1-5 w.% carbon.

CHAPTER 13

Intermediates mainly containing molybdenum, titanium and tungsten compounds

Patent (46/1) of Cabot Corporation (US) provides a process for producing
an article, composed at least partly of a carbide of molybdenum or tungsten. The composition consists of 10-62 w.% of molybdenum, while the
rest contains tungsten, carbon and impurities and corresponds to the
formula:

$$(Mo,W)C + (Mo,W)_2C$$

It is suitable for the production of, for example, filters for welding rods,
hard facing deposits and the like.

Another Patent (46/2) of Cabot Corporation describes the production of
shaped articles, consisting of at least one metal carbide, e.g. a carbide of molybdenum or tungsten and displaying the hardness and wear
resistance of these metals. The composition of the product contains
10-62 w.% molybdenum, while the rest is tungsten, carbon and some impurities. The molar ratio between the metal component and carbon is between 1.23 and 2.0. The metal carbide displays a Vickers-hardness of
at least 1800 kg/mm^2 and is suitable for the production of tubes.

Patent (51/5) of The Carborundum Company refers to the use of $MoSi_2$ as
a ceramic cement. The cement is used to join separate components of
ceramic materials, e.g. of sintered silicon carbide to form a composite
article. The components are so positioned that the surfaces to be joined,
are contiguous. Molten $MoSi_2$ is added under inert conditions at above
2030°C to fill the space between the components. Upon solidifying, a
composite ceramic article is obtained. The cement may contain a wax
binder e.g. polyethylene glycol or polyvinyl alcohol, paraffin or
methyl cellulose.

Patent (175/1) of <u>Matsushita Electric Industrial Co.Ltd. (JP)</u> relates to a resistant mass for glazings,   arranged on a carrier and fired thereon, the mass containing 6-70 w.% of molybdenum silicide in forming fine powder and 94-30 w.% of fritted glass. In addition, the mass may contain sufficient fluid carrier substance to form therewith a paste.

Patent (229/3) of <u>Shinagawa Refractories Co.Ltd. (JP)</u> concerns refractory materials, containing coarse refractory grains and, as a binder for these grains, titanium nitride, silicium nitride and aluminium oxide, formed "in situ". The coarse grains comprise: silicium carbide, mullite, refractory clay, silicium oxide ore, magnesite, magnesite-dolomite, zirconium. The titanium-containing refractory material is rutile or ilmenite. The composition may also contain graphite, electrode wastes, carbon soot and coke.

According to Patent (232/1) of <u>Sigri Elektrographit GmbH (DE)</u> refractory structural material is obtained, composed of 60-80 w.% tungsten carbide and 40-20 w.% graphite. The mixture of these components in the given proportions is shaped under pressure, whereafter the shaped body is compacted at a temperature above the peritectic point, whereby a part of the tungsten carbide smelt separates, followed by cooling of the shaped body. The tungsten carbide is applied in form of powder, with particles of $< 5 \mu m$ size.

CHAPTER 14

Intermediates mainly consisting of lithium and yttrium

In Patent (16/4) Annawerk Keramische Betriebe GmbH disclose a shaped
body, sintered and/or pressed at high temperatures from a mixture of
0.2-2.0 w.% Al_2O_3; 0.2-2.0 w.% Fe_2O_3; max. 0.4 w.% carbon, 1.0-20.0 w.%
Y_2O_3 and 76.0-98.6 w.% Si_3N_4, furthermore 0.05-3.00 w.% Li_2O (referring
to the total amount of Si_3N_4, Y_2O_3 and Li_2O). The characteristics of
the composition do not deteriorate, neither at room temperature, nor
at sintering temperature. The components are applied in such proportions
that the C-content of the final product does not exceed 0.4%.

Patent (104/10) of General Electric Company refers to a solid electrochemical element, consisting of a cell, an electrolyte, a cathode and
an anode in the cell. The anode is composed of lithium, in form of
amalgam or alloys of lithium with indium or thallium, while the cathode
is a non-stoichiometric composition of lithium, according to the formula:

$$Li_x M_y O_z$$

(x = 0-1; M = a cation the oxidation state of which depends on the value
of x; y = 0-3 and z = a value ensuring the electric neutrality of the
composition). The solid electrolyte consists of an aluminate of sodium-
lithium, placed between the anode and cathode. The structural formula
of the electrolyte is:

$$Li\ Na\ O\ 0.9\ Al_2O_3$$

containing 1.3-80% lithium ions.

Patent (176/3) of Max Planck Gesellschaft zur Förderung der Wissenschaften
e.V. concerns the production of crystalline lithium nitride, displaying
increased conductivity by heating lithium metal in an inert vessel to

40-180°C under a nitrogen pressure of at least 250 mm Hg and by raising
the temperature to the smelting point of Li_3N after the reaction has
started, or by carrying out the process at a nitrogen pressure under the
atmospheric pressure, but higher than 250 mm Hg, at a temperature higher
than 300°C. The applied lithium metal should be of 99.9% purity and
the reaction vessel in this case should be made of tungsten, niobium,
ruthenium or tantalum. Conversion has to take place in the presence
of molecular hydrogen or hydrogen bonded to one of the reacting
partners.

CHAPTER 15

Intermediates mainly consisting of silicium dioxide, silicates and ceramics, with a great variety of metal components

Patent (1/4) of Advanced Materials Engineering Ltd. (GB) describes the production of synthetic ceramic materials by forming a blend of a first component (silicium + oxygen), a second component (nitrogen and a metal differing from aluminium), this blend being heated to 1500°-2000°C in an oxygen-poor atmosphere. The first component may be silica, silicone (in the form of polymethyl-phenyl-siloxane or polymethyl siloxane) or ethyl silicate. The metal, other than aluminium, can be yttrium, magnesium, lithium.

Patent (3/1) of Agence Nationale de Valorisation de la Recherche (FR) concerns the preparation of a coating of refractory metal carbide on an article, which contains at least 0.5% carbon. The refractory material can be selected from Ti, Zr, Hf, Ta, Nb, while the carbonaceous material may consist of carbon, graphite, the carbides of tungsten, molybdenum, chrome, the alloys of cobalt and tungsten carbide, and steel. Coating takes place at 850°-1250°C in a hydrogen-containing atmosphere at atmospheric pressure during a period of 10 minutes to 20 hours.

The Anglian Water Authority (GB) present in Patent (15) a stable aqueous dispersion, containing an alkali metal silicate and an aluminium silicate (at a pH of 3-7.5) with max. 5% silicate content (expressed in terms of SiO_2), the dispersion being dissoluble in hydrochloric acid and being obtained by exposing an aqueous solution of the aforementioned components to the effect of shearing forces. The patent also provides an apparatus for the treatment by shearing forces.

In Patent (22/4) Bayer A.G. (DE) reveal a process for producing cosmo-chlore pigments according to the formula:

$$Na\ Cr\ Si\ O_6$$

The composition is obtained by heating stoichiometric (or near stoichio-metric) amounts of a component of the given formula along with mineral-ising additives, at least twice to the reaction temperature (850°-1150°C), including an intermediary disintegration. As a chrome component chrome (III)-oxide, basic chrome (III)-oxide, chrome (III) hydroxide or oxy-hydroxide or their basic salts are recommended.

Patent (44) of British Steel Corporation (GB) developed a process of manufacturing an article, having a hard phase therein, comprising the steps of infiltering a porous body, comprising one of the elements: carbon, boron or silicon, with a first liquid metallic material to coat the surface of the pores of the body, this first metallic material be-ing capable of chemically reacting with the material of the porous body to form a hard phase of the carbide boride or silicide metallic material; subjecting the body to a temperature at which the first metallic material and the material of the body react to form a hard phase of the carbide, boride or silicide of the first metallic material on the surface of the pores of the body and infiltering the body with a second liquid metallic material which does not react chemically with the material of the body substantially to fill the pores of the body. The first infiltered ma-terial contains titanium and the second copper.

Patent (50) of Le Carbone Lorraine S.A. (FR) relates to the production of synthetic carbon or graphite (of double porosity), consisting of a composition of macro-porous and micro-porous grains with an open macro-porosity, the aperture radius of the macro-pores being between 10 and 350 μm and that of the micro-pores: 0.2 - 1 μm. The basic material ap-plied can be petroleum coke, oil coke, tar residue, anthracite, soot or charcoal.

The object of Patent (51/1) of The Carborundum Company is a method for hot pressing carbon reactive refractory material into fabricated shapes, by charging the refractory material into a graphite mould, heating the material to a pre-selected sintering temperature and subjecting the

material, while at this temperature, to the action of a pressure generating means for a time sufficient to form and compact the refractory
material into a solid article, the interior surfaces of the mould and
the surfaces of the pressure generating means which contact the refractory materials being covered with a protective layer of a refractory metal
which is other than the carbon reactive element in the carbon reactive
refractory material in the case where said element is a metal. The carbon reactive refractory material comprises a carbide of silicon, boron,
zirconium, niobium, titanium, tantalum, molybdenum, or tungsten or a boride of silicon, zirconium, niobium, titanium, tantalum, molybdenum or
tungsten or a silicide of boron, zirconium, niobium, titanium, tantalum,
molybdenum or tungsten, or a nitride of boron, silicon, zirconium, niobium, titanium, tantalum, molybdenum or tungsten, or a phosphide of boron or silicon.

Patent (58/2) of the Chas.Taylor's Sons Co. (US) concerns the formation
of refractory bodies of exact final dimensions, composed of a refractory material, a sol of colloidal silicium oxide in form of an aqueous
paste, which is introduced in the cavity of a matrix of a pair of com-
plementary non-porous mould pieces, wherein the paste is cooled to
$9.4^{\circ}C$, causing an irreversible precipitation of the silicium oxide sol
between the refractory particles and thereby realising an extremely
strong bond between the components. The matrix (of a weak conductivity)
is made of an acrylic substance, teflon or a polyolefine. The final
product displays a porosity less than 25%, a density higher than 2.0 g/cm^3
and a rupture modulus higher than 84 kg/cm^2.

Patent (82/3) of E.I. Du Pont de Nemours and Company describes an amorphous aluminium silicate powder, composed of spherical particles, having
a central part of silicium oxide, aluminium silicate and one ore more
refractory metal oxides therein, this central part being surrounded by
a coating of hydrated amorphous aluminium silicate and, as the case may
be, by a surface layer over the coating, composed of a metal or a metal
oxide.

Patent (86) of Eaton Corporation (US) concerns the production of a composite material by using an intermediary "host" of porous structure of
a first material (ceramic material) and a second material, which infil-
trates the pores of the intermediary host under the effect of infiltration

forces. The second material is of a metal and is heated to at least its fusion point. The infiltration forces are left in active state, until the second(metal) material has reached its solidus.

Imperial Chemical Industries Ltd. claim in Patent (128/1) a process for the preparation of a fibre, comprising the steps of (a) fibrising a composition having a viscosity of at least 1 pois, which comprises a solvent, a metal compound dissolved in the solvent and polyethylene oxide dissolved in the solvent, wherein the proportion by weight of the metal compound is greater than the proportion by weight of poly-ethylene oxide (0.1-2 w.%) and (b) removing at least part of the sol-vent from the fibre thereby formed. The solvent is water and the metal compound is water-soluble and can be selected from the group consisting of the chlorides, sulphates, acetates, formates, hydroxides, phosphates and nitrates of aluminium, iron, zirconium, titanium, beryllium, chro-mium, magnesium, thorium, uranium, yttrium, nickel, vanadium, manganese, molybdenum, tungsten and cobalt. More particularly the metal compound is aluminium oxychloride, basic aluminium acetate, basic aluminium for-mate, zirconium oxychloride, basic zirconium acetate, basic zirconium nitrate or basic zirconium formate. The produced fibres are suitable for textiles, catalysts and for reinforcement purposes.

Patent (164/1) of Kyusyu Refractories Co.Ltd. (JP) overcomes the short-comings intrinsic in conventional unburned carbon-containing refractory bricks, i.e. oxidisation under high temperature conditions and inciden-tal decarbonisation, and the phenomenon of exfoliation and detachment of the fragile layers, by utilising a refractory brick material com-prising substantially more than 1 weight % carbon, 1-10 weight % alu-minium powder and/or magnesium powder, and optionally 0.5-6 weight % silicon powder. The carbon component may comprise a carbonaceous ma-terial like plumbago, artificial graphite, electrode waste, petroleum coke, foundry coke, carbon black, etc. The refractory brick material includes inorganic refractory material, for example general basic, neutral or acid oxides, such as magnesia, chrome, spinel, dolomite, alumina, silica and/or zircon, carbides, such as silicon carbide and/or titanium carbide, and nitrides, such as silicon nitride and/or boron nitride.

According to Patent (203) Normalair Garrett (Holdings) Ltd. and
UKAEA developed a method of preparing a refractory article by the
steps of casting a mixture, comprising a refractory material, a
binder and a non-aqueous diluent, which is compatible with the binder and is not removable during the curing step below; curing the
binder to give a moulded article, removing the binder from the moulded article and subsequently firing to form a refractory article, in
which method the diluent is removed from the moulded article before
the firing step is carried out. The refractory material used includes
mixtures of silicon carbide and carbon, furthermore materials, which
may be heat-treated to produce conventionally termed refractory material. Thus, the refractory material, which is preferably in a powdered
form, may be a glass, such as petalite or a borosilicate or lithium
alumino-silicate glass; a technical ceramic, such as porcelain or a
lithium alumino-silicate; a pure oxide ceramic such as alumina or zirconia; a carbide ceramic; a metal, such as nickel or stainless steel;
or silicon. Materials capable of being fired without shrinking are
preferred.

Patent (204/1) of Norton Company refers to producing low porosity refractory articles by blending a powdered refractory material with a
carbon-containing substance; devolatilising the blend of refractory
material and carbon-containing substance, if the carbon-containing substance contains unwanted volatiles, under such conditions as to avoid
substantial loss and polymerisation of the carbon-containing substance;
forming the blend into a shape; if necessary heat-treating the shape
so as to cause the carbon-containing substance to decompose into a
carbon residue; sintering and simultaneously impregnating the carbon-
containing refractory shape with a molten metal mixture, comprising at
least two metals, and having a coefficient of thermal expansion close
to or slightly greater than that of the refractory material, the impregnation taking place in a non-oxidising atmosphere (argon, neon,
nitrogen, hydrogen) and the temperature of the molten metal mixture
being high enough to result in uniform impregnation of the said refractory material and to bring about reaction of a substantial amount of
the molten metal with the carbon; removing the impregnated refractory
shape from contact with the molten metal mixture and cooling it. The
powdered refractory material has a numerical average particle size of

less than 350 microns and is selected from boron carbide, titanium
diboride, aluminium boride, chromium boride, silicon boride, chromium
carbide, molybdenum disilicide, zirconium boride, silicon nitride,
beryllium boride, zirconium carbide, titanium carbide, or a mixture
of two or more thereof. The carbon-containing substance is essentially
100% organic binder, selected from phenolaldehyde condensation resins,
furfuryl alcohol, furfural, and furan resins.

Patent (254/2) of Tokyo Shibaura Denki K.K. discloses a method of producing a sintered body of ceramics, wherein a powder mixture, consisting
essentially of at most 10 w.% of yttrium oxide, at most 10 w.% of aluminum oxide, at most 10 w.% of aluminum nitride, at most 5 w.% of at least
one material selected from the group consisting of titanium oxide, magnesium oxide and zirconium oxide, and the balance, essentially of alpha-
silicon-nitride, is sintered under a non-oxidising atmosphere at 1500°-
$1900^{o}C$.

Patent (254/6) of Tokyo Shibaura Denki K.K. describes a sintered body
of ceramics, comprising 0.1 to 10 w.% of Y_2O_3; 0.1 to 10 w.% of Al_2O_3;
0.1 to 10 w.% of AlN; 0.1 to 5 w.% of at least one oxide selected from
the group consisting of Li_2O, BeO, CaO, V_2O_5, MnO_2, MoO_2 and WO_3 or a
combination of at least one of these oxides with at least one oxide
selected from the group consisting of B_2O_3, MgO, TiO_2, Cr_2O_3, CoO, NiO,
ZrO_2, Nb_2O_5, HfO_2 and Ta_2O_5; and the balance of Si_3N_4. The process
for producing a sintered body of ceramics comprises: molding a powder
mixture of the same composition, and sintering the resultant molded
compact in a non-oxidative atmosphere. The process requires no hot
press and therefore is very suitable for bulk production.

Another Patent (254/7) of Tokyo Shibaura Denki K.K. refers to producing
a sintered body of ceramics, comprising 0.1 to 10 w.% of yttrium oxide;
0.1 to 10 w.% of aluminium oxide; 0.1 to 10 w.% of aluminum nitride;
0.1 to 5 w.% of at least one silicide, selected from the group consisting
of magnesium silicide, calcium silicide, titanium silicide, vanadium
silicide, chromium silicide, manganese silicide, zirconium silicide, niobium silicide, molybdenum silicide, tantalum silicide and tungsten sili-
cide; and the balance being silicon nitride. The process for producing

a sintered body of ceramics takes place according to Patent (254/6).
The described process yields products of high density and impact strength.

Patent (257) of <u>Toyo Boseki K.K. (JP)</u> reveals a process for the preparation
of a metal carbide-containing molded product, which comprises heating a
molded composition, comprising at least one powdery metal selected from
the group consisting of B, Ti, Si, Zr, Hf, V, Nb, Ta, Mo, W, Cr, Fe and
U and having an average particle size of not more than about 50 μ and
an acrylonitrile polymer at a temperature of 200 to 400°C and then calcining the resulting product at a temperature of 900 to 2500°C in an
inert atmosphere. The heating of the molded composition in form of
fibres or films is carried out in an oxidising atmosphere (air). The
particle size of the powdery metal is not more than about 10 μ. In
this process, the molded composition can be produced with ease in a continuous state, and the treatment for its stretching can be effected with
good workability to increase the strength. By preparing the molded composition in such continuous form, the continuous supply thereof into a
furnace for heat-treatment is made possible, without any special apparatus, and the workability at the heat treatment steps is excellent to
afford the objective metal carbide-containing molded product in a continuous form such as filaments or a film. Thus, the production of the
metal carbide-containing molded product is attained simply in the combination of a conventional installation for preparation of a molded
product of an acrylonitrile polymer and an installation for carbonisation.

CHAPTER 16

Treatments for materials, e.g. carbon, graphite and other
organic or inorganic substances for influencing the properties
of refractories and ceramics (for example coating)

Patent (3/2) of Agence Nationale de Valorisation de la Recherche concerns
the production of powders, suitable for sintering by forming powder particles or droplets from an aqueous salt solution, wherein the cation/anion
ratio is established according to the material the droplets are made of,
the droplets being suddenly cooled to minus 70°C, causing their instant-
aneous freezing, followed by lyophilisation for reducing their water content, the mass obtained being further heated in order to develope the
final crystalline structure. The droplets of the salt solution display
a size between 50 and 300 μ.

Allied Eneabba Ltd. (Australia) developed in Patent (8) a process for
treating zirconium, comprising the coating of zirconium grains with a
strong basic reagent; calcining the coated grains and processing the
calcined grains in order to remove the calcined impurities. The strong
basic reagent is sodium hydroxide or sodium carbonate. The amount of
sodium hydroxide is 2-4 w.% of zirconium. Calcining takes place at
750°-800°C.

Asahi Glass Co.Ltd. claim in Patent (19/5) a process for producing shaped
bodies of a ceramic material or metal, through mixing a ceramic or metal
powder with a resin that can be dissolved in an organic solvent and a
resin that cannot be dissolved in an organic solvent, the shaped product
being processed with an organic solvent for dissolving the resin, which
is dissoluble in an organic solvent, whereafter the product is heated
to burn out the rest of the resin.

ASEA A.B. disclose in Patent (20/6) a process for producing ceramic articles by isostatic compression of a blank, from ceramic or metallic powder, with the aid of a gaseous pressure agent. The blank is placed in a casing of glass or vitrifiable material, while the casing is made gas-impermeable prior to the isostatic pressure process. The article and casing are placed in a vessel, resistant to sintering temperature, during sintering the casing being transformed in a fused element, the surface of which is limited by the vessel wall, the blank being arranged over this surface. The pressure necessary for the isostatic process is established by the gaseous agent on the fused element.

Patent (22/2) of Bayer A.G. refers to a process for obtaining carbon foam, on the basis of carboimide group containing foam, by heating such foam at a rate of $100^{o}C$ up to $2000^{o}C$ per hour in an inert atmosphere and holding the product (carbon foam) at a temperature between 600° to $3000^{o}C$ for a period of time from 30 minutes to 50 hours. Patent (22/13) of Bayer A.G. refers to a similar process, using instead of carboimide groups containing foam a foam which comprises isocyanate groups.

Patent (31/1) of D. Bitzer (DE) is concerned with the production of a carbonaceous matrix, which had been reinforced by whiskers or fibres, by mixing together whiskers, fibres, a fine grained carbon material, polymers, inorganic or pre-polymerised organic materials, the mixture being sedimented in a fluid, from which the fluid phase is separated and sedimented reinforced matrix is shaped under heat and pressure and, as the case may be, coked or graphitised. The reinforcing elements form 5-80% of the carbonaceous material.

D. Bitzer presents in another Patent (31/2) a process for producing a carbon-rich layer on metallic or ceramic surfaces, the carbon-rich layer being applied to such surfaces and being treated in an oxidising atmosphere, at $350^{o}C$. The carbon-rich organic materials are mixed with prepressed, expanded graphite with a density of 0.25-1.85 g/cm^{3}.

Yet another Patent (31/3) of D. Bitzer concerns the production of carbon coatings on carbon or metal compositions, by applying non-ionic, cationic or anionic acrylamid polymerisates on the material to be coated,

followed by pyrolysing the applied layer at a temperature between 110°
and 200°C. The pyrolysis of the polymerisates on the substrate to be
coated can be carried out step-wise at various temperatures between
110° and 220°C in the presence of a gas current.

Patent (41) of British Leyland UK Ltd.(GB) relates to the production
of shaped bodies of sintered metal compositions, from an initial mixture of a metal powder, a polybutyraldehyde binder and a plasticising
agent (dibutylphthalate), this mixture being shaped into a tube, which
then is hardened at 180°-220°C and sintered at 1280°-1300°C. To the initial mixture a lubricating agent (stearic acid, stearin) is added for
promoting injection molding.

Patent (62/1) of Chemotronics International Inc. (US) concerns a method
of preparing an uncracked carbon structure by infusing at least one
strand of a flexible polyurethane resin with a liquid thermosetting
polymerisable furan resin or furan resin precursor, so as to cause the
strand to swell by infusion of the liquid furan resin or furan resin
precursor into the strand; removing substantially all the non-infused
furan resin or furan resin precursor from the surface of the strand;
polymerising the furan resin or furan resin precursor with a catalyst
and carbonising the polymerised furan resin or polymerised resin precursor infused polyurethane structure under inert or reducing or vacuum
conditions in less than 5 hours at a rate of temperature change which
would cause cracking of the strand if substantial amounts of thermosetting furan resin or furan resin precursor remained on the surface and
which reproduces the shape of the original polyurethane strand.

Patent (62/2) of Chemotronics International Inc. provides a carbon
structure, comprising at least one crack-free strand of carbon, which
reproduces the shape of a polyurethane resin strand, infused with a
liquid thermoset furan resin or furan resin precursor, substantially
free of a surface coating of liquid furan resin or furan resin precursor and then cured so as to thermoset, and with a cured infused weight
of up to six times that of the polyurethane strand alone and which
during and after carbonisation of the polymerised furan resin or polymerised resin precursor infused polyurethane structure resists cracking

when rapidly heated to 2500°C under inert or reducing or vacuum conditions. The polyurethane resin strand is a strand in a reticulated
polyurethane resin structure, the carbon structure having a density
of from 0.03 to 0.08 gm/cc. The polyurethane resin strand may be a
strand in a polyurethane resin fibre structure.

According to Patent (62/4) of Chemotronics International Inc. a
reticulated structure is produced by applying a thermosetting foam,
composed of interconnected cells, divided by membranes. These membranes have a section, which is smaller than the fibres forming at the
intersections of the membranes. The cells are interconnected by ope-
nings in the membranes in order to permit the introduction of a com-
bustible gas. The gas is ignited for destroying the cells' membranes,
thereby realising a reticulated resinous structure, which then is carbonised.

Patent (91) of Elettrocarbonium S.p.A. (IT) refers to a process for
coating the surface of presintered carbon articles by applying a glassy
layer of boron acid anhydride, which impedes oxidation. The process
consists of preparing pre-sintered carbon article, preparing an aqueous
solution of diammonium penta-borate, applying this solution to the surface of the carbon article, followed by drying the solution. The coating
can be applied by immersion, by brushing or by spraying.

Patent (111) of GR-Stein Refractories Ltd. (GB) provides a ramming mix,
composed of a refractory material comprising 55 - 90 w.% calcined anthracite grains, up to 40% graphite, 3-25 w.% of a thermo-setting resin and
4-15 w.% of an alcoholic wetting agent e.g. monoethylene glycol. The
ramming mix is suitable for the linings of aluminium electrolysis cells
and eliminates the disadvantages of prior art methods, in which the
ramming mix, comprising carbonaceous aggregates and a tar/pitch binder
has to be installed hot by a labour-intensive procedure under considerable fume emission.

Patent (141) of Inoue-Japax Research Inc. (JP) refers to a body consisting of carbon with at least a superficial portion, that is more
graphitised than the rest of the body. The more graphitised portions

display a "p" value between 0.3 and 0.6, determined by x-ray diffraction and on the basis of the formula:

$$d = 3.440 - 0.086 \ (1-p^2)$$

where: d = the average stratic spacing of the structure of turbostratic carbon obtained by an x-ray diffraction diagram. Graphiti-
sation is effected by passing a continuous electric current through
superficial portions of the body, this current being superimposed by
analternating high frequency current.

Patent (150) of Kernforschungsanlage Jülich GmbH,(DE) concerns a cast
body, provided with corrosion resistant coating, the body consisting
of particles of graphite or particles of synthetic graphite or similar
substances and a binder. The body is supplied with one or more coating
layers, of the material the body is made of, but which also contains
silicium or zirconium in such a manner that the layer nearest to the
body has the lowest silicium or zirconium content, while the exterior
layer contains the largest amounts of these additives. After coating,
the body is coked under a protecting gas atmosphere between 650° and
850°C, whereafter it is exposed to a temperature between 1550° and
1800°C (at a very high rate) in order to develop silicium carbide in
the external layer(s).

Patent (157) of V.I.Kostikov et al.(USSR) reveals the production of
antifriction elements by heating a carbon blank in an inert gas atmosphere (or in vacuum) to 1800°-2200°C, the fluid smelt being impregnated
with silicium and then cooled. Impregnation takes place first at 2100°
to 2200°C during 3-5 minutes and thereafter at 1800°-2050°C during
30-40 minutes, followed by cooling to 200-150°C at a rate of 50-100°C/min.
The carbon blank has an open porosity of 25-50% and contains particles
of 30-120/um size.

Patent (161) of Kureha Kagaku Kogyo K.K.(JP) provides a method of producing shaped articles of carbon, wherein a shaped article made of a
vinylic high polymer containing chlorine and having an atomic ratio,
R, of chlorine atom to carbon atom in the range

$$\frac{2}{7} < R < 1,$$

is first treated in an infusibilising step, at a temperature below 400°C
with at least one substance selected from ammonia, amines and a mixture
of these, and then the treated shaped article is subjected to a carbonisation step in a non-oxidising gas atmosphere. The shaped article,
after having been carbonised, is further subjected to a graphitising
treatment, furthermore to an additional treatment of stretching the product, carried out during the infusibilising step of the method. The
final carbonised or graphitised product is additionally coated with a
substance deposited thereon selected from boron, carbon (graphite), and
silicon carbide.

Porous ceramic shaped body, mainly of silicium, can be obtained according
to Patent (185/5) of Motoren- und Turbinen-Union München GmbH which is
coated with a material of composition Ag_2O (refractory metal . O_x), or
the mixture of such materials, using as refractory material tungsten,
molybdenum or tantalum. A suspension is made of fine silver powder and
oxide(s) of refractory metals, which then is applied to the shaped body
by immersion, brushing or spraying.

Patent (190/3) of NGK Insulators Ltd. relates to a process for producing
silicium containing, non-oxide, refractory materials by heating a mixture
of a Si-powder, along with a thermoplastic resin, the components being
kneaded together and the obtained product being injection-molded. The
thermoplastic resin is an organosilicium polymer, which is introduced
into the refractory ceramic material under dissociation of the polymer.

Patent (190/9) of NGK Spark Plug Co.Ltd.(JP) concerns a process for
producing high density sinter articles, by forming from an inorganic
powder and an organic resin a composite body, through press molding or
injection molding, the resin then being burnt out from the obtained
blank, the surface of the blank being coated with latex rubber layer,
whereafter the blank is still more compressed, isostatically, under a
pressure of 2 t/cm^2 and then sintered.

Patent (201/1) of NL Industries Co. provides a fire retardant, smoke-
suppressing agent in halogen-containing polymer compositions, comprising
a magnesium-zinc complex salt of an acid, prepared by reacting magnesium

oxide with a zinc salt of an acid in a molar ratio of from 2:1 to 10:1.
The zinc salt is a salt of an inorganic acid. The zinc salt can be selected
from zinc sulphate, zinc borate, zinc phosphate, zinc phosphite, zinc
fluoborate, zinc sulphite, zinc fluo-silicate, zinc silicate, zinc sulpha-
mate, zinc trimellitate, zinc pyromellitate, zinc terephthalate, zinc
fumarate, zinc phthalate, zinc maleate, zinc silicylate, zinc gluconate,
zinc tartrate, zinc isophthalate, zinc orthophthalate, zinc adipate,
zinc glutamate, zinc lactate and mixtures thereof.

In Patent (216) <u>Ramu International (US)</u> claim a process of freezing an
inorganic particulate slurry or suspension containing a freeze-sensitive
colloidal ceramic sol, which comprises supercooling the slurry in a
freezing media (a hydrophobic liquid) and then freezing the slurry to
cause spontaneous nucleation, resulting in the formation of a very large
number of icecrystals that are consequently very small thus producing
a ceramic structure that is very uniform throughout. According to
one aspect, the invention includes the addition of lithium ions to the
freezing media with or without supercooling. Also composite ceramic
structures, composed of a laminate of at least two different fired ceramic
bodies, in which at least one ceramic body material has a different coefficient of expansion from at least one other wherein each ceramic body
are described. The slurry is a refractory slurry and the refractory is
alumina, magnesia, zirconia, silica, zircon, mullite, uranium oxide or
clay. The particulate slurry may contain lithium ions in a sufficient
amount to inhibit ice crystal growth.

Patent (218/1) of The Foundation: <u>The Research Institute for Special</u>
<u>Inorganic Materials (JP)</u> relates to a process for producing a thermal-
resistant ceramic fired body by adding an organosilicon polymer as an
additive to at least one ceramic powder selected from oxides, carbides,
nitrides, borides and silicides; molding the resulting mixture in a desired form and then firing the molded body at a temperature of 800° to
$2000^{\circ}C$ under vacuum or in at least one atmosphere selected from inert
gas, carbon monoxide gas, carbon dioxide gas, hydrogen gas, nitrogen
gas and hydrocarbon gas, using as additive at least one organosilicon
polymer selected from organo-polysiloxanes, organosiloxane copolymers,
polycarbosiloxanes and nitrogen-containing organosilicon polymers.

Schunk & Ebe GmbH (DE) present in Patent (225/1) a carbon boat for
aluminium evaporation, the surface of the boat being coated (completely or partly) with a refractory protective coating, composed of titanium and/or zirconium and/or hafnium and/or tantalum or the nitrides
thereof. The coating is applied by flame or plasma-spraying or by
chemical gas dissociation. The metal coating is exposed to a nitrogen
or nitrogeneous atmosphere at $800^{\circ}-1400^{\circ}C$ and sintered.

According to another Patent (225/2) of Schunk & Ebe a smooth carbon body
with improved gliding properties at high temperature can be produced by
doping it with boron or aluminium, the green body of the product being
calcined and, as the case may be, graphitised.

Siemens A.G. (DE) developed in Patent (231) a process for producing
ceramic products like clinker, floor tiles, building elements, fine stone
ware, electric and thermal insulation elements, porous products, these
products being made of the inorganic residue forming during the low-
temperature carbonisation of oil shale. The inorganic residue is diluted
with finely dispersed inorganic and/or organic additives or plasticising
agents.

Patent (232/2) of Sigri Elektrographit GmbH concerns a protective layer,
made of metallic materials on the surface of carbon or graphite electrodes,
the protective layer containing fibrous insertions in an amount of 0.1-10
vol.%. The surface of the insertion materials has to undergo a chemical
or mechanical treatment. The inserted fibres are, as a rule, tangentially
oriented to the electrode surface or they can be placed in a net-like
arrangement.

Sigri Elektrographit GmbH also disclose in Patent (232/4) a carbon-containing contact mass for the production of an electric bond of high stability between high-temperature elements. The cold-processable mass consists of a mixture of epoxide resin, tar as a binder and graphite and
metal powder as the solid components. Prior to application, the mass is
mixed with an acid-free hardening agent, for example polyamine. The
electric and thermal conductivity of the contact mass corresponds to
those of carbon blocks and is suitable for bonding such blocks with me-

tal conductors, for example anodes and cathodes of cells for aluminium
fusion electrolysis.

Patent (238/2) of Société Européenne de Propulsion S.A. (FR) refers to
the gluing of refractory bodies. More particularly the patent concerns
porous refractory bodies or which can be made porous by thermal treatment.
The gluing is carried out by placing between the bodies a layer of 0.1-1mm
thickness of a glue that leaves behind, after thermolysis, a porous refractory residue, and charges, composed of refractory particles of 1-150 μu
size, thermally dissociated in such a way that a porous joint forms between porous refractory bodies. The combination of refractory body and
joint is densified by chemical precipitation (in vapour phase) of pyrocarbon or a ceramic material.

Another Patent (238/3) of Société Européenne de Propulsion S.A. concerns
a composite structure, obtained by densifying a fibrous texture of a
ceramic material, other than carbon, by means of a refractory matrix,
filling up the initial open porosity of the fibrous texture and which
contains an oxide of aluminium or an oxide of zirconium (with a fusion
point over 1750°C). Densification takes place by chemical sedimentation,
in the gas phase, under conditions permitting the diffusion of the
reactive gas mixture at the base of the pores prior to sedimentation.

Société Française des Pétroles BP (FR) reveal in Patent (239/1) the
production of carbon and graphite foams, by exposing pitch foams to
oxidation, the foams processed then being carbonised and, as the case
may be, graphitised. The initial foamy substances can be residues of
the vapour cracking process of asphalt, of oil and their mixtures.

In Patent (239/2) Société Française des Pétroles BP describe a process
for producing rigid foams by expanding an initial tar mass under the
effect of a pore forming agent, which can be selected from p,p'-oxybis-
benzene sulfonyl hydrazide; azodicarbonamide (or azobisformamide);
azo-1,1'-cyano"-1-cyclohexane; azobisisobutyronitrile; diazoamino-
benzene;N,N'-dimethyl-N,N'-dinitroso-terephthalamide; N,N'-dinitroso-
pentamethylenetetramine; benzene sulfonyl hydrazide; 4-toluene sulfonyl
hydrazide, diphenylsulfone-1,3-disulfonyl hydrazide; p-toluene sulfonyl
semicarbazide; 5-phenyl tetrazole, diphenoxy-4,4-disulfo-hydrazide tri-
hydrazinotriazine.

Patent (243/2) of Süddeutsche Kalkstickstoff Werke A.G. (DE) relates to carbon bodies, obtained by compressing finely dispersed carbon along with 3-50 w.% Si O$_2$ dust and 5-25 w.% water under a pressure of from 150 kg/cm^2. The SiO$_2$-dust contains particles of 10-50 µu. The process advantageously uses wastes or by-products. For example, the SiO$_2$-component is made of filter dust, isolated from the waste gases of silicium or ferrosilicium treatments, while the carbon component is composed of the lime-carbon by-product, isolated in processing calcium cyanamide, the major part of such carbon consisting of graphite. The carbon bodies produced can be used as reduction agents in electrothermal processes.

According to Patent (263/1) and essentially Patent (263/2) of UBE Industries Ltd. (JP) fine metallic nitride powders, having a high purity, are prepared without causing any plugging and other troubles in the reaction apparatus and with easy heat control of the reaction, by reacting a metallic halide with liquid ammonia to form a metallic amide or metallic imide, separating the resulting metallic amide or metallic imide from the reaction mixture and thermally decomposing the separated product in an atmosphere of nitrogen or ammonia to produce metallic nitride powder. The reaction is carried out in the presence of an organic solvent having a specific gravity higher than that of the liquid ammonia, the organic solvent and the liquid ammonia being not or only slightly soluble with each other at a reaction temperature, and the metal halide is allowed to react with the liquid ammonia by introducing it into the lower organic solvent layer of the reaction system, which separates into two layers of the upper liquid ammonia layer and the lower organic solvent layer due to the difference of the specific gravities. The metallic halide is selected from SiCl$_4$, BCl$_3$, TiCl$_4$, VCl$_4$, SiBr$_4$, TiBr$_4$, GeCl$_4$, HSiCl$_3$ and H$_2$SiCl$_2$, and the organic solvent is selected from n-pentane, n-hexane, n-heptane, cyclohexane, cyclooctane, benzene, toluene, xylene and any mixtures thereof.

Union Carbide Corporation present in Patent (265/3) a process for producing graphite bodies with low longitudinal thermal dilatation from a mixture, that contains coke, a carbonisable thermoplastic binder. The bodies are formed by extrusion, the extrudates being carbonised and graphitised to obtain the final graphite body. The process applies coke, obtained from coal tar of such type that under calm conditions a continuous mesaphase is forming.

Patent (267/21) of UKAEA is concerned with a process for producing a
porous body by preparing fibres of tar, by integrating the fibre(s) into
an initial mass, so that the fibres are in contact with one another, this
mass then being treated to carbonise the fibres (having a ∅ of 5-200
microns). During heating the surface of the fibres is softened to such
an extent, that they are fused together at the contact places.

In Patent (269) United States Borax and Chemical Corporation (US) provide
an improved oxidation-resistant material by a method, which comprises:
pressing a mixture comprising (a) 40-60 w.% of carbon, (b) 10-40 w.%
of silicon, titanium disilicide, boron silicide or silicon carbide or a
mixture of two or more of these and (c) 10-40 w.% of boron carbide,
boron silicide, titanium diboride, zirconium diboride or boron nitride
or a mixture of two or more of these, while maintaining the mixture at
a temperature efficient to densify it. The mixture is subjected to a
pressure of 1.0 - 1.5 tons/square inch, at a temperature of $1700^{o}-2200^{o}C$,
to establish therein a density at least 80% of the theoretical.

Patent (272/3) of Varta Batterie A.G. (DE) reveals a process for producing ceramic lithium nitride, suitable for use as a solid electrolyte
in galvanic lithium cells. Li_3N-powder is mixed with molten Li metal,
which acts as a binder, in a noble gas atmosphere. After cooling, the
mixture is compressed into a shaped body and then sintered at $550^{o}-750^{o}C$
in a nitrogen atmosphere. During sintering the Li-binder will be converted into Li_3N as well. Due to the bond with lithium, it is possible
to expose the pulverulent initial material to rolling (milling) treatment. From the obtained ductile rolled product pellets of exact dimensions can be produced.

The object of Patent (273) of VE Wissenschaftlich-Technischer Betrieb
Keramik (GDR) is a production process for sintering adjuvants of low
expandibility and high resistance to temperature changes, furthermore of
chamotte bricks, on the basis of cordierite-forming substances and
chamotte, the said adjuvants consisting of: 40,0-60,0% kaolin chamotte;
14.5-9.5% crude magnesite; 23.3-15.5% crude clay II.R.; 12.7-8.5%
"Vetro"-clay and 9.5-6.5% calcined alumina. Heating and cooling are
effected under closely controlled conditions. Cordierite is introduced

into the composition prior to cooling, for effecting devitrification, thereby producing a fine crystalline structure with a glass phase < 5%.

Vereinigte Grossalmeroder Thonwerke and KWM Keramik-Werk Mering GmbH & Co. KG (DE) disclose in Patent (275/1) sintering adjuvants with improved resistance to temperature changes, using as raw material chamotte, cordierite, sillimanite, SiC or their combination with quartz glass, applied in an amount of 5-50 w.%. The glass quartz consists of particles of 0-7 mm.

Patent (281) of M.I. Kolesnik et al. (USSR) provides a process for protecting against oxidation the carbon-containing parts of metallurgical aggregates by impregnating the said parts with orthophosphoric acid to form thereby a glassy mass in the pores of the said part. The impregnated part is dried and then once again impregnated with anSi- or Ti- or Al-base composition, capable of forming the phosphate, followed by heating to $300^{\circ}-350^{\circ}C$.

In Patent (284/1) Westinghouse Electric Corporation (US) disclose a method of producing ceramic-forming oxide powders with variable and predetermined sintering characteristics such as MgO, BaO, Ta_2O_5, ZnO, CeO, SnO, CaO, Cr_2O_3, CoO, As_2O_3, Sb_2O_3, ZrO_2, TiO_2, ThO_2, Y_2O_3, SiO_2, UO_2, Al_2O_3, and TiO_2. There is a relationship between sintering rate of oxide powders produced from metal alkoxides and degree and nature of hydrolysis. This relationship between the degree of hydrolysis of the corresponding alkoxide and the density of the resulting oxide is determined by hydrolysing several samples of the alkoxide using different amounts of water in diluted systems. The samples are sintered to produce the metal oxide and its density is measured. A desired density can be achieved by using the amount of water of hydrolysis which corresponds to that density. Oxide powders with very high rate of sintering characteristics can be achieved by using at least 5 moles of water of hydrolysis per mole of alkoxide. The relationship between density and temperature is represented by the following diagram:

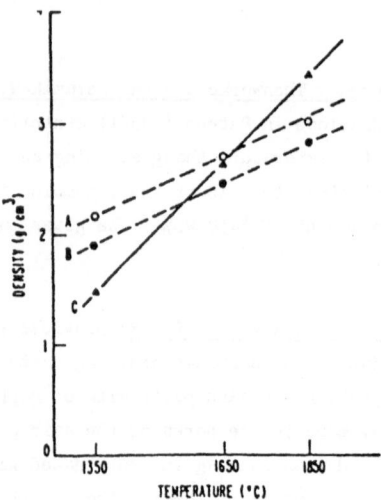

<u>Zirconal Processes Ltd.</u> developed a process (287/5) for producing shaped
refractory elements by mixing together a refractory powder with a gel-
forming liquid binder for obtaining a fluid dough-like mass, which then
is placed in a mould, wherein it hardens into a green body, this green
body then being dried and calcined. The liquid binder consists of an
aqueous salt of zirconium, which contains a gel-forming component,
(for example amino-alcohol, morpholin, calcined magnesium powder,
magnesium spinel).

CHAPTER 17

Ceramic intermediates based on natural clay and minerals

Patent (2) of Advanced Mineral Research A.B. (SE) refers to a material
having a substantially fibre- or flake-like microstructure and is manufactured from a starting material, comprising mainly natural or synthetic silicate mineral, having a hardness exceeding the value 6 on the
Mohs-hardness scale and containing on the one hand SiO_2, and on the other
at least one of the oxides MgO, FeO, Al_2O_3 and Fe_2O_3, by finely dividing
this material to a specific surface area of at least 15,000 cm^2/cm^3,
measured according to Blaine, and subsequently hydrothermally treating
the resultant powderous starting material at a temperature of approxima-
tely 175-325°C and at a pH of at least 9 in the presence of water or
condensed steam. The starting material may contain natural or synthetic
olivine or granate mineral.

Patent (7) of Albright & Wilson Ltd. (GB) concerns compositions for road
surfaces, consisting of 40-75 w.% of a spinel; 20-45% of a non-calcined
clay and 1-10 w.% of a precalcined aluminium silicate (calcined at min.
800°C). The spinel is composed of mineral chrome or a residue obtained
in the chemical extraction of chromates from mineral chrome.

Patent (29) of Bellfires B.V., Hapert N.V. Koninklijke Sphinx (NL) relates
to the production of the rear wall of a crack-free, open fire place, according to which a refractory material is pressed into a matrix, followed
by burning and cooling it. The refractory material consists of cordierite,
and mullite. In the rear wall one or more fissure(s) are made, which
then are filled up with a mixture, containing the aforementioned components, furthermore sodium silicate glass (1 w.p.), the filling mixture
then being hardened. In a rear wall of a width of 600-700 mm the fissures
made therein should be 2-10 mm wide and 10-25 cm long.

According to a Patent (47) of R. Campbell Stein (GB) a mineral mixture is provided for use in manufacturing a refractory material, comprising an admixture of zircon and andalusite. Andalusite is an alumino-silicate mineral found in nature in particle sizes from about 10 mm and less and frequently 3 mm and less. Thus, andalusite as occurring in nature, frequently contains a high proportion of particles of a size which are immediately suitable for producing a well balanced packing without the need for further crushing and screening. Typically andalusite contains 50-60% alumina, the remainder being substantially silica and some minor impurities, which are fusible at calcination temperatures and thus can act as a flux. Andalusite has the advantage that its natural grain has a low porosity and when calcined, its crystal form is changed but it does not suffer from undesirable change in volume. The material is suitable for fabricating refractory products (bricks, monoliths, etc.) by moulding into shape the mineral mixture (zircon, andalusite) with or without the addition of a binder and firing the moulded product.

Coordination et Développement de l'Innovation S.A. (CORDI), (FR) disclose in Patent (69) a process for producing ceramic articles by applying, between the refractory particles, a synthetic feldspathic substance, obtained in the compression phase by a reaction between a clayey material and an alkaline agent, during a few minutes, at 100°-200°C, the feldspathic substance forming 1-5 w.% of the refractory material, which can be used in the structure of ceramic oven feed elements and as cores in metal smelting equipment.

Patent (70/5) of Corning Glass Works refers to the production of crystal-containing gels and papers, films, fibres, boards and coatings made therefrom. The process for making the gels comprises three general steps: first, a fully or predominantly crystalline body is formed containing crystals consisting essentially of a lithium and/or sodium water-swelling mica, selected from the group of fluorhectorite, hydroxyl hectorite, boron fluorphlogopite, hydroxyl boron phlogopite, and solid solutions among those and between those and other structurally-compatible species selected from the group of talc, fluortalc, polylithionite, fluorpolylithionite, phlogopite, and fluorphlogopite; second, the body is contacted with a polar liquid, desirably water, to cause swelling and disintegration of the body, accompanied with the formation of a gel, and third, the solid

liquid ratio of the gel is adjusted to a desired value depending upon
the application thereof. Where papers, films, fibres, boards or coatings
are desired, these are prepared from the gel, and to impart good chemical durability thereto, are thereafter contacted with a source of large
cations to effect an ion exchange reaction between the large cations and
the Li^+ and/or Na^+ ions from the interlayer of the crystals, and the
products then dried. Glass-ceramic bodies are the preferred starting
materials.

Patent (81/4) of Dresser Industries, Inc. (US)describes carbon-bonded
refractory shaped articles with 1-35 w.% carbon, the rest being composed
of a non-basic refractory aggregate, 75% of which consists of andalusite.
A feature of theproduct is that after degasification no shrinkage takes
place at about $1100^{o}C$.

Fayenceries de Sarreguemines Digoin & Vitry-le-François S.A. (FR)
developed in Patent (92) a vitrifiable material (sandstone, porcelain,
fayance) of high resistance in non-worked state and capable of enduring
thermal changes of great amplitude and speed. The vitrifiable material
consists of a natural mineral (30. w.%) which contains lithium (petalite,
spodumene) and 0.5-5 w.% of an organic resin (polyvinyl acetate; a copolymer of vinyl acetate and various non-saturated ethylene monomers).
The ceramic composition is sintered at $1230-1290^{o}C$.

Patent (105/1) of General Refractories Company (US) is concerned with
the production of refractory cement, composed of a mixture of chamotte
of cordierite, obtained by reaction sintering, in very fine form and with
a low thermal expansion coefficient. More particularly the chamotte
contains: 12-15 w.% MgO; 33-37 w.% Al_2O_3 and 45-51 w.% SiO_2, the
aggregate weight of these components forming 96.5% of the total weight
of the composition, which also comprises a binder, e.g. a solution of
colloidal silicium or aluminium, furthermore methylcellulose and a
plasticising agent (clay "hydrite MP").

The object of Patent (127) of Motoyuki Imai (JP) is a sol of particles
comprising: montmorillonite, or natural or synthetic hectorite, tetra-
silicic mica or taeniolite, these particles having been made hydrophobic

by treatment with a titanic acid ester, a zirconic acid ester, a silane, having at least one methoxy-, ethoxy- or silanol-radical and at least one vinyl-, epoxy-, acrylic- or amino-radical, a beta-diketone mixed with lauryl amine, a titanium amide, a zirconium amide, or a cationised silicone oil. The patent also provides a method for making a heat-resistant, electrical insulation of the said sol by dispersing mont-morillonite, or natural or synthetic hectorite, tetra-silicic mica, or taeniolite in water, whereby these materials are swollen and cleaved into a sol, mixing this sol with a solution in an organic solvent of a zir-conic acid ester, a titanic acid ester, a silane having at least one methoxy-, ethoxy or silanol-radical and at least one vinyl, epoxy-, acrylic-, or amino-radical, a beta-diketone mixed with lauryl amine, a titanium amide, a zirconium amide or a cationised silicone oil, sepa-rating the cohered material formed thereby, washing it with an organic solvent, drying the material to at least 95% solids, dispersing the solid as a dilute solution in an organic solvent, forming the mixture into a layer and blending this mixture with a solution of a resin in the same organic solvent, forming the blend into a layer and removing the solvent.

Imperial College of Science & Technology (GB) present in Patent (129) a method of obtaining clay-like materials from naturally occurring ig-neous rocks. Natural swelling clays have a number of uses, for example as binders for foundry sand and in pelletising, as dispersants, as fil-lers in paints and plastics. The patent provides synthetic materials, which can be used as substitutes for these clays. According to the patent there is provided a process in which an igneous rock is subjected to a hydrothermal treatment by being agitated and heated in the presence of water to a temperature of from 350 to 450°C under a pressure of up to 350 atmospheres. The material which results from this hydrothermal treat-ment has properties which make it similar to the naturally occurring swelling clays. Thus, the product has saponite-like properties, e.g. its X-ray diffraction analysis is consistent with saponite although it does not necessarily have the same structural state. For convenience saponite can be represented by a planar unit cell of the structural formula:

$$\left[Si_{8-x}Al_x\right]^{IV}\left[Mg_{6-x}Al_y\right]^{VI}O_{20}(OH)_4.$$
$$\searrow M_{x-y}$$

In this formula m_{x-y} indicates the presence of compensator cations to-
talling x-y valencies per unit cell, with x-y roughly equal to unity, while
the other symbols have their usual significance.

Patents (158/1) and essentially (158/2) of Hans Kramer GmbH (DE) refer
to a mineral composition on the basis of a reversibly expandable, large
surface, crystalline triple-layer mineral, with separate crystalline
layers, essentially composed of hydro-mica and vermiculite of fine
particles. The mineral may also contain intercrystalline, embedded,
non-expandable components.

S.A. Lafarge (FR) reveal in Patent (166/4) a composition for the production of insulating bricks, by casting from a mixture that contains:
67-75% of chamotte (with 42-44% aluminium content, of 6000 cm^2/g finesse),
19-29% of calcium carbonate (with a 6500 cm^2/g finesse), 3% of a hydraulic
binder (aluminous cement) and 1% of a fibrous substance (silico-aluminous
mineral fibres). The mixture also contains (less than 1%) secondary additives, for modifying the rheology (e.g. sodium polyphosphate methyl cellulose).

In Patent (172/2) Magnesital-Feuerfest GmbH (DE) disclose calcined blocks,
composed of 70-93 w.% MgO-Al_2O_3 spinel, 2-8 w.% Al_2O_3, 1-9 w.%   binding agent and up to 27 w.% of refractory additives. The spinel is present in the form of fusible grains. The additives contain 2-13.5 w.%
of $CaZrO_3$, while the binding agent contains 0.3-3 w.% of clay. The mass
composed of the aforementioned components is formed into blocks under a
pressure over 1000 bars and sintered at 1650°-1850°C. The blocks are
resistant, not only to heat, but also to complex chemical effects (both
oxidising and reducing effects).

In Patent (177) O.Mayer (DE) claims ceramic sintered shaped bodies, containing 10-80 w.% of glass and 20-90 w.% of materials of volcanic origin,
like basalt, diabase or slag and/or quartz sand and/or pumice, furthermore up to 1 w.% of inorganic activating substances like silicium
carbide or soot. The components are mixed together in finely dispersed, dry state, shaped and sintered at 750-1500°C. The shaped
product displays a thermal expansion coefficient of about 70×10^{-7}
mm/mm$^\circ$C (at 800°C).

NL Industries, Inc. provide in Patent (201/5) a process for producing
a composition, which contains: smectite-type trioctahedral clay, which
in turn contains oxides of Li, Na, Mg and Si (expressed in terms of
oxides) and F, according to the formula:

$$a Li_2O : b Na_2O : c MgO : dF : 8SiO_2$$

wherein: $0,25 \leq a < 1,1$, $0 \leq b < 0$, 0, $4,75 < c < 5,85$, $0,5 < d \leq 3,5$,
$0,6 \leq a+b < 1,25$, $6,0 < a+b+c < 6,65$,

or: $0 \leq a < 1,2$, $0 \leq b < 0,60$, $4,75 \leq c < 7,0$, $0 \leq d < 4$, $0 \leq a+b < 1,4$
$5,5 < a+b+c < 8$,

and $c \geq$ in case $a + b = 0$

An aqueous suspension is prepared, containing 12-35 w.% of solid substances according to the indicated chemical compositions, while combining magnesium oxide, water, hydrofluoric acid, furthermore a base
selected from lithium hydroxide or sodium hydroxide (or their mixture)
until the base is dissolved, whereafter silicium oxide sol is added to
the suspension, followed by hydrothermal reaction of the gel and drying
of the composition.

The object of Patent (214/1) of Quarzwerke GmbH (DE) is a material fo
manufacturing porous high-quality ceramic, which comprises, as a raw
material anorthosite,which, if heated for the first time from room
temperature up to about 1000^oC, has a permanent expansion of at least
about 1%, the anorthosite being present to compensate the shrinkage in
a clay-containing material containing raw materials which naturally
tend to shrink. The anorthosite may be of the kind which has been
saussuritised by metamorphosis and/or metasomatosis and which has a
permanent expansion of about 3.2% if heated for the first time up to
1000^oC. The anorthosite is present in a quantitative proportion of
from 5 to 95 w.%.

Another Patent (214/2) of Quartzwerke GmbH also provides fine ceramics,
suitable for example for wall tiles. The ceramic mass again consists
mainly of anorthosite and the production process is similar to that as
described in Patent(214/1). Anorthosite is applied in crude or pre-
sintered state, in combination with foamed blast furnace slag.

According to Patent (215) of <u>Quigly Co.Inc. (US)</u> a high purity magnesium aluminate spinel can be obtained by mixing finely dispersed aluminium oxide with a source of magnesium (magnesium hydroxide) at a
0.4-0.8 w. ratic , followed by heating the mixture to $900^{\circ}C$ during a
time, necessary for forming the spinel. The mixture also contains a
a fluxing agent (fluor, cryolite, aluminium fluoride).

In Patent (264) <u>Unibra S.A. (Be)</u> present novel intercalation compositions of kaolin, consisting of an organic radical, selected from ammonium salts of carboxylic acids (with more than 2 C atoms), alkaline
metal salts of carboxylic acids (with more than 2 C atoms), glycol
alkylene according to the formula:

$$HO-CH_2-(CH_2)_n-CH_2-OH$$

(wherein n = 0 to 4) and quaternary ammonium radicals, for example according to the formula:

$$\begin{pmatrix} R_4 & & R_1 \\ & N & \\ R_3 & & R_2 \end{pmatrix}$$

wherein: R_1, R_2, R_3 and R_4 = each a group selected from aliphatic or
aromatic hydrocarbons or stabilised derivates thereof.

According to Patent (287/2) of <u>Zirconal Processes Ltd.</u> a mullite is
synthesised by reacting in the presence of water stochiometric quantities of a lower alkyl silicate and a water soluble aluminium compound,
capable of being hydrolysed to a polymerisable substance e.g. aluminium
chlorohydrate or aluminium alkoxide. The reaction product contains
silicon,oxygen and aluminium and on firing gives a residue containing
the mullite. The most useful industrial application of the invention
is to cause or allow the reaction system to gel into a rigid and
coherent mass,into which refractory grains can be incorporated and
of which nozzles can be produced.

Patent (252/2) of <u>Tokuyama Soda K.D. (JP)</u> concerns the production of
a material of high absorption capacity for liquids. The material consists of a fibrous substance and a mineral powder of lamellar particles,
containing pores with radii smaller than 0.5 μm, in a volume of at
least 2.5 cm^3/g. The mineral powder is composed of, for example, calcium
silicate, displaying a gyrolite structure and a molar ratio of SiO_2/CaO
between 1.6 and 6.5. The powder also may contain a sulphate of calcium.

CHAPTER 18

Special treatments
(by electron bombardment, electro-arc, explosion, irradiation, etc.)

Degussa A.G. (DE) developed in Patent (73/3) a process for controlling
the reaction of gases with solid substances, mainly the nitriding of
shaped silicium bodies. The control takes place on the basis of a predetermined temperature/time programme, through the adjustment of pressure and at the same time measuring the amount of consumed reaction
quantities. A limit value transmitter establishes the optimum maximum
pressure and maximum feed. As soon as the maximum charge for maintaining the maximum pressure is achieved, the limit value transmitter stops
the programmed process and the pressure adjustment and commands the
introduction of an amount of the reaction gases reduced by 10-50%.

Patent (93) granted to A.Yi-Hung Fang (US) refers to the recovery of
phosphate from acid waste solutions, forming during some aluminium
treatments and containing considerable amounts of dissolved aluminium
phosphate and nitric acid. The recovery takes place by a reaction of
the aqueous, acid waste solution, with sodium hydroxide or sodium carbonate in such a way that nearly all of the aluminium phosphate is transformed into sodium aluminate and sodium tri-phosphate, while saltpetre
is transformed into sodium nitrate and phosphoric acid into sodium
tri-phosphate, which then is recovered by crystallisation.

Gaz de France and S.A.Fonderies Gailly (FR) describe in Patent (103) a
process for separation from a particulate mass, the volatile or com-
bustible substances therein. According to this process, the mass is
introduced in a flame, thereby separating the fragments of the mass from
one another. The air amount necessary forproducing the flame and the
combustion temperature of the mass to be treated are so adjusted that
no chemical change occurs in the purified mass. The process can be used
in removing oil from metal cuttings (chips).

Patent (132/2) of Institut Khimicheskoi Fiziki Adademii Nauk SSSR
(USSR) reveal to production of diamonds and/or various compositions of
diamond structure, made of boron nitride, from an initial material like
carbon and/or boron nitride, which can be converted under the effect of
explosive energy, produced by detonating a charge, which contains an
explosive and the material to be converted. According to the patent,
explosives are used, ensuring during explosion dynamic pressures between
3 and 60 GPa, and temperatures between 2000 and 6000°K. As explosives
cyclotrimethylenetrinitroamine, cyclotetramethylenetetranitroamine,
trinitrotoluene, trinitrophenylmethylnitroamine, tetranitroerythrol,
tetranitromethane and their mixtures are recommended. The charge may
also contain inert additives (1-50 w.%), which evaporate or decompose
ahead of the explosion wave. Such inert additives may contain liquid
nitrogen, aqueous metal salt solutions, hydrated crystals, salts of
ammonium, hydrazine and the like.

Patent (165) of Laboratoires OM S.A. (CH) relates to a pharmaceutical
composition of anti-acid action and high adsorbing capacity, which contains, as active components, magnesium silicate and aluminium silicate,
furthermore polydimethylsiloxane. The composition is processed into
tablets, powder or granules.

According to Patent (187/1) of National Aeronatic and Space Administration
(NASA)   polycrystalline ceramic bodies of aluminium oxide, silicium
carbide, silicium nitride, can be made more resistant to shocks by coating
them with a layer of polycrystalline ceramic material, displaying a low
elasticity modulus and having micro-fissures in their interior. The individual grains of the composing materials present such an anisotropy of
thermal expansion coefficients, that micro-fissures develop therin under
heating. The material of low elasticity modulus also includes phases of
different volumes, which cause the formation of micro-fissures by the
volume changes taking place during the phase transformations in the
material.

Nipkti po CHerna Metalurgia (Bulgaria) disclose in Patent (192) a
method of producing a layer of protective coating on a cylindrical
carbon article by means of electric arc treatment of a layer material
on the surface of the article, in which the electric arc continuously

burns between the material on the article to be treated and a lateral
electrode, while the surface of the article moves helically relative
to the lateral electrode, wherein the electric arc is stabilised in a
magnetic field and the electric arc treatment is carried out at a
current above 600 A and wherein the helical movement has a pitch larger than 10 mm and produces a helical band on the article having a
width of more than 10 mm when measured in a direction parallel to the
axis of the article. If the carbon article is to be used as an electrode for an electrothermal furnace, it is made with a plurality of
layers of protective coating of high aluminium content, which layers
of coating are produced by the above method and each possesses a helical band created by the electro-arc treatment with a width of more
than 10 mm measured in a direction parallel to the axis of the electrode.

The object of Patent (209/2) of The Plessey Co.Ltd.   is a method
of manufacturing an article of substantially pure dense ceramic material (silicon nitride), which includes the following three stages:
firstly, mixing a powder of the substantially pure ceramic material
with an additive which promotes densification and is capable of nuclear
transmutation into a gas when exposed to radiation, and hot pressing
the mixture to form a billet of dense ceramic material in which at
least some of the additive survives as an impurity; secondly, irra-
diating the billet to convert the surviving additive by nuclear transmutation into a gas, which is held captive in the billet; and thirdly,
subjecting the billet to a hot forging operation during which the captive gas escapes and an article of substantially pure dense ceramic
material is forged from the billet. The additive is provided by lithium,
or beryllium, or compounds of lithium or beryllium. When the additive
is lithium or its compounds, the quantity which is mixed with the silicon nitride is such that, prior to irradiation, lithium is present in
the dense billet to the extent of at least 5 parts per million and not
more than 5 per cent by weight. The irradiation may take the form of
bombardment by protons or bombardment by neutrons. In either event the
lithium is transmitted into helium. Helium, being a gas of low atomic
weight, is readily absorbed into the silicon nitride, in which its
atoms are temporarily held captive.

Patent (241) of R. Stolz (DE) aims at improving the strength of aluminium oxide by chilling with an emulsion consisting of oil and water, in which various solid materials are dissolved.

According to Patent (255/2) of Tokyo Shibaura Co.Ltd. ceramic bodies of nearly theoretical density can be produced by preparing from a ceramic powder a green body of high porosity, which then is pre-sintered into a body of less than 30% porosity, and which is then embedded into a pressure transmitting powder. Next pressure is applied to the embedded green body at 1600°-1900°C, for uniformly compressing the pre-sintered body and for removing the pores therefrom.

PART III

Finished products and components made of
refractory and ceramic materials

This part deals with the production of electric, electronic, mechanical and other parts from refractory and ceramic materials, in so far as these parts have not been described in Part II of this study.

It is often difficult to distinguish between an intermediate and a finished product. For example, abrasive material could be considered as being an intermediate, but when it is processed into a grinding wheel it is, without any doubt, a finished product.

Part III covers only finished products with the exception of some catalyst materials and intermediates which are to be used for the production of bricks for various purposes, such as furnace walls.

CHAPTER 19

Electric, electronic and ion-conducting products and components

19.1   Electric devices   and components

Patent (39/4) of Robert Bosch GmbH refers to the production of cermet
electrodes for gas sensors with ion-conducting solid electrolytes.
The electrodes are composed of finely dispersed ceramic material for
building up a supporting structure (for example of zirconium dioxide,
thorium oxide, hafnium oxide and/or cerium oxide). The ceramic material the supporting structure is made of is less sinter-active than
the ceramic material of the solid electrolyte.

Patent (62/3) of Chemotronics International Inc. (US) reveals the formation
of carbon elements, according to which, in a first phase, at least one
thin filament of elastic polyurethane resin is infused by a polymerisable
fluid furane resin (or a precursor thereof) in such a manner that the
filament will be expanded, followed by the removal of the non-infused
resin (or precursor) from the filament surface (second phase);   the
polymerisation of the precursor by a catalyst (third phase) and the carbonisation of the formed element (fourth phase) in a neutral or reducing
atmosphere, under vacuum, in such a manner that the temperature changes
take place at a rate, causing the fissuring of the filament, when essen-
tial amount of thermosetting resin remained on the filament surface.

In Patent (80) Draloric Electronic GmbH (DE) disclose a process for producing polycrystalline ceramic cold conductors, which are partly n-con-
ductive and partly p-conductive and which consist of a ferroelectric
material of perowskite structure according to the formula:

$$A^{2+} B^{4+} O_3$$

with doping substances. A^{2+} indicates barium and at least one of the
metals: strontium, potassium, lead, while B^{4+} consists of titanium (in
excess amount) and for shifting the curie-temperature also of zirconium
or tin.

Patent (99) of Fordath Ltd.(GB) relates to a method of manufacturing or
treating a spark-erosion electrode, comprising a shaped block of substantially pure graphite, including the step of pyrolytically depositing
onto the outer layers of the block, carbon formed by the thermal decomposition of a gaseous carbon-containing material under controlled temperature and pressure; by heating the machined block in a moving stream
of a gaseous carbon-containing material at a steady temperature in the
range $650^{\circ}C$ to $1500^{\circ}C$, at which decomposition takes place, without any
substantial pressure difference existing across the block and under controlled conditions of flow rate and partial pressure. The pyrolytic
deposition takes place for a period between 15 and 40 hours in a furnace,
from which all the air has been purged by gaseous nitrogen or a non-
decomposing volatile organic material, prior to introduction of
gaseous carbon-containing material,which is to be thermally decomposed.
The proposed electrodes are capable of more efficient metal removal
than has hitherto been the case with graphite electrodes.

Patent (124/7) of Hitachi Ltd. concerns a silicon carbide electrical
insulator material of low dielectric constant,comprising an electrically
insulating sintered body with silicon carbide as a main constituent and
an element providing electrical insulating properties in sintered state,
wherein the sintered body comprises an element selected from Group V of
the P.S. and the carrier concentration within the crystal grain of silicon carbide is 5×10^{17} cm^3 or less. The relationship between the
specific dielectric constant   carrier concentration is illustrated by
the following diagram:

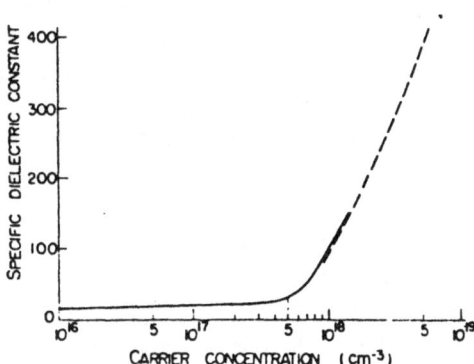

19.2   Electronic components

Patent (57) of <u>Champion Spark Plug Company (US)</u> provides a sintered
semi-conductor body, comprising from 45 to 65 w.% of silicon carbide,
having a median particle size of from 6 to 12 microns and substantially
all of which has a particle size of from 1 to 25 microns in size; from
8 to 15 w.% of alumina, from 5 to 15 w.% of calcium oxide and/or magnesium oxide, and the balance of silica, the semi-conductor body having
an apparent porosity of from 10 to 25 percent. This semi-conductor
body is useful in a jet engine igniter of the high energy type, which,
in service, is fired by an ignition system including a capacitor. The semi-
conductor body is incorporated in the high energy igniter so that a
portion of a surface thereof is adjacent a spark gap between a centre
electrode and a ground electrode. When voltage is applied to the
centre electrode, there is a limited flow of current along the semi-
conductor surface. This current flow causes ionisation of gas in the
spark gap. The ionisation enables a spark discharge to occur at a lower

voltage than would be required without the ionisation. Discharge of
the previously charged capacitor, which occurs when there is a spark
between the ground and centre electrodes is responsible for the high
energy nature of the spark.

Fujitsu Ltd. (JP) developed in Patent (101) a ceramic substrate, consisting essentially of (based on the weight of the substrate) 0.5 ⌐ 5.0 w.%
of MgO and 95.0 ⌐ 99.5 w.% of the total of Al_2O_3 and SiO_2, the proportion
of the Al_2O_3 to the SiO_2 being in the range from 50:50 to 80:20 by weight
and wherein the grain boundaries between mullite crystals ($3Al_2O_3.2SiO_2$)
are filled with cordierite ($2MgO.2Al_2O_3.5SiO_2$). The ceramic substrate
can be used as a circuit base for mounting an integrated circuit chip.

Patent (124/4) of Hitachi Ltd. describes a high-temperature thermistor,
composed of a polycrystalline sinter-body, containing a composition,
selected from Be, BeO, Be_2C, B, BN, B_2O_3, B_4C in an amount of 0.1-8 w.% ,
the rest consisting of SiC, the inevitable impurities containing 0.1 w.%
Al, 0.2 w.% Fe, 1 w.% Si and 0.4 w.% free carbon. The thermistor is
composed of a pair of electrodes, arranged on the surface of the polycrystalline sinter body, and wires, connected to the corresponding
electrode.

Rosenthal Technik A.G. (DE) present in Patent (222) an alumina porcelain
composition, for use, after firing, in forming electric insulators, consisting of 40 to 65 w.% of calcined alumina, 20 to 40 w.% of plastic
kaolinite-montmorillonite constituents, and 20 to 26 w.% of a flux combination, which contains alkali-aluminium silicates and 0.1 to 5 w.% of
alkaline earth oxides in the form of alkaline earth compounds, e.g.
barium oxide in the form of barium compounds (thus avoiding the disadvantages associated with the use of TiO_2 and MnO_2.

Stettner & Co. (DE) developed a ceramic sinter body in Patent (240),
suitable for use as ceramic insulator in electro-technical devices, on
the basis of mineral olivine, the body containing in sintered state:
10-30 w.% MgO + FeO; 15-40 w.% Al_2O_3; 40-65 w.% SiO_2, the FeO content
being 3-7%. During sintering a crystalline phase of magnesium-ferrocor-
dierite is forming. The ceramic body may also contain 4 w.% of an alkali
oxide, e.g. K_2O.

According to Patent (244) of <u>Sulzer Frères S.A. (CH)</u> electric heating elements are composed of at least three layers of ceramic material, whereby the resistance of at least one of the layers is reduced by doping with regard to that of the other layers, the doped layer determining the electric parameters of the element, the second layer preventing the diffusion of atoms, molecules, into the doped layer, while the geometrical dimensions of the non-doped third layer determine the thickness. The doped and non-doped layers are made of different ceramic materials.

Patent (256/3) of <u>Toshiba Ceramics Co. Ltd. (JP)</u> refers to elements of devices for producing semi-conductors, obtained by the precipitation of a silicium carbide layer on a carbon substrate, this layer displaying an x-ray diffraction peak of the (200) plane with a half-width of 0.35° or less, measured by C_u-K_{α} -radiation, used in x-ray diffraction analyses. More than 30% of the polished surface of the silicium carbide layer is composed of silicium carbide crystal grains, having a max. width of more than 0.15 t + 5/u (t = the thickness of the silicium carbide layer in/u). From this material crucibles for separating silicium monocrystals or susceptors or heating elements or reaction tubes can be made.

19.3   Ion-conducting components, suitable for use in batteries, electrolytic cells, fuel cells, solid electrolytes

<u>Aluminium Company of America (US)</u> presents in Patent (10) a mixture, which can be used in electrolytic cells for producing aluminium for the formation of impermeable welds. The mixture contains essentially carbon, a binder and a solvent (boiling point: 150-350°C), consisting of an aromatic hydrocarbon (methyl-naphthalene) with saturated side chains and a boiling point between 150° and 350°C, the solvent being used in an amount of 8-16% of the binder's weight. The mixture composition displays high electric conductivity.

Patent (11/1) of <u>Aluminium Pechiney (FR)</u> relates to the refractory coating of electrolysis vats, produced by restituting old coatings by a process, comprising the coating of the bottom and walls of the vat with a heat-insulating bed, made of a mixture of C, NaF, CaF_2, Al_2O_3,

Na_2SO_4, $CaSO_4$, while the cathode, arranged on the heat-insulating bed consists of a pulpy substance of C, NaF, CaF_2, Al_2O_3, AlF_3, Na_2O and a binder composed of residues of petroleum and oil.

Battelle Institut e.V. (DE) disclose in Patent (26) a galvanic element with a negative electrode, consisting of an`alkali metal (sodium) which is solid at the operative temperature, and with a positive electrode of chlorine or bromine, while the electrolyte (which is fluid at the operative temperature) consists of aluminium chloride, sodium chloride, sulphur dioxide, the galvanic element also containing a separator of beta-aluminum oxide (of fine crystals) of identical size, arranged on a porous support.

According to Patent (39/1) of Robert Bosch GmbH solid electrolyte tubes for the sensitive element of measuring devices, defining the oxygen content in exhaust gases, can be made of stabilised zirconium dioxide. One end of the tube is closed, this end part of the tube being made of zirconium dioxide, stabilised by yttrium oxide and/or ytterbium oxide, while the rest of the tube consists of zirconium dioxide, stabilised by calcium oxide. The closed end tube is provided with a small plate of stabilised zirconium dioxide, while the tube end may also be provided with a cup-shaped element of zirconium dioxide, stabilised by Y_2O_3 and/or Yb_2O_3.

Another Patent (39/5) of Robert Bosch GmbH also describes a solid electrolyte made of zirconium dioxide ceramics, this solid electrolyte comprising, at least partly, an electrode layer or a system of several layers, of which the layer nearest to the solid electrolyte is an electrode layer, the arrangement also comprising an intermediary layer, which is placed between the electrode layer and the layer system on the one hand, while on the other hand, being attached to the solid electrolyte and displaying a higher ion-conductivity than the solid electrolyte. The intermediary layer consists of fully stabilised ZrO_2 and contains Y_2O_3 and/or Yb_2O_3 and/or Sc_2O_3 as a stabilising oxide.

Yet another Patent (39/6) of Robert Bosch GmbH refers to a porous ceramic material, consisting of zirconium dioxide and 20-60 w.% titaniumdioxide. The ceramic material can be used as a diffusion-resistant or protective layer of electro-chemical primary elements.

In Patent (48) of <u>Canadian Patents and Development Ltd. (Ca)</u> a method
is revealed for producing a ceramic hydrogen ion conductor. More in
particular a polymorph, which has a high conductivity for protons (H^+)
in the form of hydronium ions (H_3O^+) is achievable with the compound
known as $H_3O^+ - \beta'' - Al_2O_3$ via the direct production of a dense
$K^+ - \beta'' - Al_2O_3$ material with a high percentage (80% or more by weight)of
the β "-alumina phase . That phase is ion exchangeable with the hydronium ion. Such potassium β "-alumina can be produced by using as
constituent feed components sodium, potassium, magnesium and aluminium,
freezing them ,then freeze-drying and calcinating them. The resultant
is $K^+ - \beta'' - Al_2O_3$ in an ultra-fine white powder form. The potassium ion
(K^+) of the compound is readily ion exchangeable with H_3O^+.

<u>Chloride Silent Power Ltd. (GB)</u> developed various types ceramic mate-
rials used in electrochemical or other energy conversion devices.
Thus, for example, Patent (63/1) of this firm provides a process for
producing beta alumina ceramic material, mainly in tubular form by
compressing powdered beta alumina or a mixture of powdered materials
which, on heating, produces beta alumina and sintering the compressed
powder material to form an impervious ceramic material, the article,
after sintering, being pressurised in a gaseous pressure medium at an
elevated temperature ,below but within, 500^oC of the sintering temperature. Preferably the temperature is within 400° of the sintering temperature.

Another Patent (63/2) of <u>Chloride Silent Power Ltd.</u> also refers to the
production of beta-alumina articles by a process comprising: (<u>a</u>) compacting finely-divided particles of beta-alumina or a mixture of finely-
divided powdered material which react together, on heating, to form beta-
alumina, so as to produce a homogeneous green compact, and sintering the
green compact to form an impervious polycrystalline ceramic; (<u>b</u>) isostatically pressurising the sintered body at a temperature between 1200^oC
and 1550^oC at a pressure above 5000 p.s.i; (<u>c</u>) cooling this body under
pressure to a temperature below 1200^oC; and (<u>d</u>) then releasing the
pressure. The process yields an impervious fine-grained polycrystalline
ceramic of uniform density in excess of 3.2 g.cm^{-3}, which can be used in
sodium sulphur cells and other electrochemical devices requiring the
passage of sodium ions.

The object of Patent (63/5) of <u>Chloride Silent Power Ltd.</u>is to provide
an improved solid ionic conducting component for joining to metal or
other electronic conductors and a method of manufacture of such components. According to this patent, there is provided a component of solid
ionic conductive material in which, over part of the component, the bulk
properties of the ionic conductor are modified by the substitution of
ions which remain firmly bonded in the material for the ions for which
the material is conductive, whereby the bulk properties in the modified
part of the component differ from the bulk properties in the unmodified
part of the component. In the case of beta-alumina, in the required
region sodium ions may be replaced by ions of a bivalent material,for
example,barium or calcium. More in particular, a beta-alumina ceramic
solid ionic conductive material is produced in which powdered material
comprising beta-alumina or a mixture which when heated will produce
beta-alumina is formed into a green shape and then fired. The composition of the powder is modified in a part of the component in order to
reduce the ionic conductivity in that part either by the inclusion of
barium or calcium or barium oxide or calcium oxide in making up the
material for part of the green shape or by doping the green shape with
a barium or calcium-containing material.

A beta-aluminium oxide ceramic electrolyte material is also disclosed
in Patent (63/6) and essentially in Patent (63/9) of <u>Chloride Silent
Power Ltd.</u> ,comprising sodium as cationic material, furthermore a ceramic
material of 8.4-9.0 w.% Na_2O; 0.6-1.4 w.% MgO; 0.3-0.7 w.% Li_2O, the
rest being composed of Al_2O_3. A sodium-sulphur cell, containing solid
electrolyte material with such beta-aluminium oxide therein, is recommended for separating fluid sodium from fluid sulphur/polysulphides.

Patent (63/12) of <u>Chloride Silent Power Ltd.</u> provides beta-alumina
electrolyte material for use in a sodium sulphur cell, protected on the
side exposed to sodium, by a thin coating of an ionically conductive
material,which will conduct sodium ions. The coating may be,for example,
a glass ,such as a sodium oxide silica glass or a sodium oxide phosphoric
oxide glass. The material may be formed by calcining a sodium-doped sol
or sol-gel applied to the surface, for example,a sodium-doped alumina sol.
Another technique comprises applying a solution or dispersion of a suitable sodium compound such as sodium tetraborate or sodium silicate or
sodium phosphate, to the surface and firing after evaporating the solvent.

Patent (64) granted to <u>V.P.Chviruk, et al.(USSR)</u> relates to graphite
packing material for alkali metal amalgam decomposers and to a method of
producing them. The material is intended for use in electrolyte processes as a means of decomposing alkali metal/mercury amalgam obtained in
the cathode cells in the course of the manufacture of alkalis .The graphite
packing material is prepared by mixing together a carbonaceous base material of petroleum or shale coke or graphite, a carbon-containing binder, comprising coal pitch, petroleum pitch or phenol - formaldehyde
resin and from 24.8 to 52 w.% of titanium in the form of titanium metal,
titanium oxide or titanium carbide, moulding the mixture into shapes,
firing the shapes at a temperature of from 500 to 1200°C and graphitising the fired shapes at a temperature of from 1700 to 1800°C.

<u>The Commonwealth Scientific and Industrial Research Organisation
(Australia)</u> developed in Patent (65/2) a solid mixed electrolyte material, consisting of a mixture of at least one non-electrolytic component (phase) and at least one component (phase) displaying a high
oxygen ion conductivity. The microstructure of the material comprises
an intimate blend of fine grains of the conductive and non-electrolytic
components, the grains of the conductive component making up 25-75 vol.%
of the blend. The non-electrolytic component contains aluminium oxide,
aluminous porcelain or mullite.

Patent (66/2) of <u>S.A. Compagnie Générale d'Électricité</u> concerns the
sintering of tubular ceramic parts. More particularly, in forming a
solid beta-sodium alumina electrolyte tube, which is disposed in a
sintering chamber, an extra thickness is imparted to one end of the
outer wall of the tube, the tubular part being inserted in an opening
provided in a plate which is disposed inside the chamber so that the
part is kept suspended in the chamber by means of the extra thickness
which rests against the edge of the said opening, and a solid part which
included a body and a shoulder that bears against the edges of said end,
is inserted in the tubular part; the chamber then being brought to the
sintering temperature. The ratio between the length of the extra thickness and the whole length of the tubular part lies between 0.1% and 5%.

The tube is illustrated in the following figure:

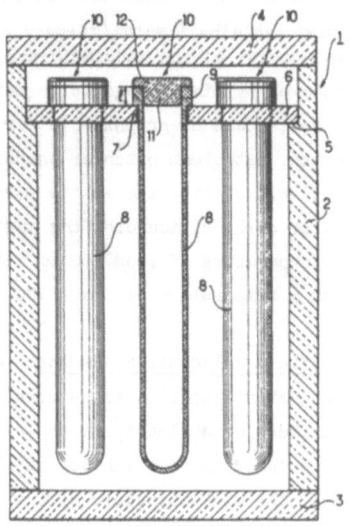

Patent (66/15) of <u>Compagnie Générale d'Électricité</u> refers to an alkaline aluminium oxide composition, suitable for use in sodium/sulphur
electrochemical cells. The composition contains at least an oxide of
a transition element (iron, cobalt, nickel, manganese) in an amount of
between 10 and 1000 w.parts per million with regard to the total weight
of the composition.

Another Patent (66/16) of <u>Compagnie Générale d'Électricité</u> describes the
continuous production of an intimate mixture of aluminium oxide and
sodium carbonate by realising an Al_2O_3/Na_2O ratio between 5 and 11,
heating the mixture in an open crucible, followed by cooling and crushing
the powder thus obtained, which then is shaped and sintered. During
sintering an additive is added to the mixture, capable of being decomposed, thereby forming a thin layer of glass (of 33 Na_2O, 67 B_2O_3,
12.5 Na_2O, 75 B_2O_3, 12.5 Al_2O_3, 27.5 Na_2O, 27.5 B_2O_3, 45 SiO_2 compositions)
on the product , which displays ionic conductivity with regard to sodium
ions. The product obtained is recommended for use in sodium-sulphur gene-
rators for electric traction.

According to Patent (67) of the <u>Commissariat à l'Énergie Atomique (FR)</u> solid electrolyte elements for fuel cells are produced on the basis of stabilised zirconium, displaying improved conductivity due to the addition of Al_2O_3. The process consists of preparing an alcoholic solution of a zirconium salt, of a salt of a metal stabilising the zirconium, like yttrium and an aluminium salt.From this solution a co-precipitate is formed of zirconium hydroxide, the said metal's hydroxide and aluminium hydroxide. The precipitate is then washed in a hydrophilic solvent for removing the traces of water, followed by drying and calcination (in air). From this material the solid electrolyte elements are obtained by compression and sintering.

In Patent (176/4) the <u>Max Planck Gesellschaft zur Förderung der Wissenschaften (DE)</u> disclose compositions of the general formula:

$$Me_{1+x-y} (H^+ \text{ resp. } H_3O^+)_y Zr_2Si_xP_{3-x}O_{12}$$

where: x = a number higher than 0-3; y = a number higher than 0-1+x and Me = a monovalent metal, for example Na, K and/or Li. The composition described is suitable for use as a solid electrolyte, conducting H^+ and/or H_3O^+ and is applicable to electrochemical cells.

The <u>Moskovskii Khimiko-Tekhnologicheskii Institut Imeni D.I.Mendeleeva (USSR)</u> claim in Patent (184/1) a process for manufacturing articles from a powder charge,based on hexagonal boron nitride with a disordered structure, which comprises mixing the hexagonal boron nitride with 1-5 w.% of the charge, of a dope capable of forming a liquid phase and volatilising during high-temperature sintering, moulding this charge into blanks, sintering the blanks in an oxidising medium at a temperature of from 1600 to 1800°C, so that the liquid phase volatilises, the charge being under a two-layer burden with an outer layer,consisting of a carbonaceous material, and an inner layer in which the material is comprised of a refractory material inert to boron nitride at the sintering temperatures. The powder charge may contain 1-5 w.% of graphite,hexagonal boron nitride. The dope is at least one oxide selected from oxides of magnesium, calcium, yttrium and lanthanum, or is at least one borate selected from borates of magnesium, calcium, yttrium and lanthanum. Articles manufactured in accordance with the patent can be used in electronics, radio engineering, chemistry, and metallurgy, as a refractory material, as electrical insulator or dielectrics in electronic instruments, as crucible material for smelting metals and aggressive alloys, or anti-friction parts.

Patent (87) of Ebauches S.A. (CH) refers to the production of lithium nitride in a compact polycrystalline structure by sintering. Lithium nitride is dissociated into powder and then the powder mass is placed in a piston-type matrix of a material, which is stable with regard to lithium nitride. Then the mass is exposed to pressure of about 2000 kg/cm^2, at a temperature between 400° and $815^\circ C$. Sintering takes place, as a rule, in a neutral (nitrogen) atmosphere. The compact poly-crystalline composition is suitable for use in lithium ion conductors, electronic insulators, solid electrolyte batteries, in measuring devices of lithium ion concentration or nitrogen concentration.

Patent (190/4)of NGK Insulators concerns solid electrolytes, which can be obtained from ZrO_2 ceramic material, consisting of cubic crystals and monocline crystals and containing Y_2O_3. The average particle size of monocrystals is not more than $2\,\mu m$, the ratio of the intensity of the monocline (iii)-x-ray radiation scattering line to the the cubic (200) line being 0.01 to 2.5.

N.V. Philips Gloeilampenfabrieken (NL) present in Patent (207/1) ceramic multi-layer capacitors, having intermediate electrodes, containing non-noble metals, such as nickel and cobalt and are made from ceramic masses, comprising alkaline earth zirconates, wherein up to 10 mole% of the zirconium may have been replaced by titanium; iron or nickel or manganese being added as doping agents to the stoichiometrical perows-kite basic compound,having a composition of $E(Zr_{1-x}Ti_x)O_3$, where $0 < x \leq 0.07$ and E is an alkaline earth metal; and the whole assembly is sintered in a reducing atmosphere.

Patent (259) of Toyota Jidosha Kogyo K.K. (JP) provides a stabilised zirconium oxide for use in oxygen-ion conducting solid electrolytes, the zirconium oxide also containing scandium oxide and ytterbium oxide according to the formula:

$$\alpha ZrO_2 \cdot \beta Sc_2O_3 \cdot \gamma Yb_2O_3$$

wherein: α, β, γ = mole fractions and $\alpha + \beta + \gamma = 1$.

Patent (207/16) of UKAEA contains improvements in or relating to the
preparation of ion conductors. More particularly the patent recommends
the preparation of a material of the general formula:

$$Na_{1+x}Zr_2Si_xP_{3-x}O_{12}$$

which is used as an ion conductor, and is obtained by preparing a precursor for the material by forming a precipitate, containing zirconia
(or a precursor therefor) and at least some of the other elements appropriate to the material. For materials with the highest conductivities
for applications as solid electrolytes with low resistance, x preferably
takes a value in the range: $1.8 \leqslant x \leqslant 2.2$.

Patent (272/2) of Varta Batterie A.G. (DE) concerns a mixed crystal
according to the formula:

$$Na_{1+ax}Zr_{2+2/3x-ax}Si_xP_{3-x}O_{12-2/3x},$$

wherein: $0.8 \leqslant a \leqslant 0.9$ and $1.8 \leqslant x \leqslant 2.3$
The crystalline solid solution displays an excellent Na-ion conductivity,
mainly in case of constant $a = 0.88$ and constant $x = 2.2$, which makes
its use in Na/S cells attractive. The high density of the single phase
monocline crystallising material, which obtains values as high as
3.10 g/cm^3, is achieved by high-temperature (1200°C) sintering in pure
oxygen atmosphere.

Patent (284/2) granted to Westinghouse Electric Corp. refers to the
production of beta-alumina ceramic material. More particularly, according to the patent a liquid polymer is formed in a low temperature
polymerisation reaction, involving organo-metallic sodium and aluminium compounds, at least one of which is partially hydrolysed. The
polymer is hydrolysed, dried to form an amorphous sodium beta-alumina
precursor, and then heated, at between 1200°C and about 1550°C to form
a ceramic, comprising ion-conductive sodium beta-alumina which is useful
as a solid electrolyte.

CHAPTER 20

Gas turbine parts, engine parts, machine tools, parts of
blast furnaces and various steel foundry equipment and
other applications

20.1   Gas turbine parts

According to Patent (75) of Deutsche Forschungs- und Versuchsanstalt
für Luft- und Raumfahrt e.V. (DE) a process for the hot isostatic pressing of porous bodies of complicate shape and made of a ceramic material,
comprises the following phases: the shaped porous body is embedded in
a press powder; the press powder present in the body is compacted by a
cold-isostatic process; the pre-compacted product of the preceding process is smelted in glass capsules. The shaped body consists of silicium
nitride, the press powder of boron nitride. The material described is
suitable for use in the construction of turbine blades.

In Patent (66/14) Compagnie Générale d'Electricité reveal a process for
producing a dense ceramic product of silicium carbide,by heating above
1200°C, under vacuum, a mixture of pulverulent alpha-phase silicium carbide and pulverulent boron, followed by sintering of the mixture at
2100°C in an argon atmosphere. The product can be used in the production
of turbine elements.

According to Patent (276/2) of Volkswagen A.G. (DE) gas turbine parts
are produced from ceramic material. More particularly,the radial rim
part (carrying the blades) of the gas turbine is produced by reaction
sintering, while the disc part is produced by hot pressing, in two separate processes, these parts being fitted together at their contact
faces by a refractory cement, thereby obtaining a monolithic structure.
The material is suitable for use in gas turbine parts, operating under
high temperature.

Patent (255/4) of <u>Tokyo Shibaura Electric Co.Ltd.</u> concerns a monolithic
composite body of a ceramic material of several pre-shaped components,
having a density of 98% of the nominal density, a bending strength of
more than 50 kg/cm^2 at 1200°C. The joining places of the composition
elements contain no foreign matter. The composite body can be shaped
into highly complicated forms, for example with three-dimensionally
curbed surface and can be used in the structure of turbine rotors.

In Patent (204/6) of <u>Norton Co.</u> a turbine wheel is described of caramic
materials. The turbine wheel consists of a hub part and a blade part,
composed of silicium nitride or silicium carbide or their mixtures, the
blade part being coated with an impermeable layer of a refractory material, selected from silicium carbide and silicium nitride. The hub part
and blade part are united with one another through diffusion bonding.
The spec. weight of the hub part amounts to 90-100% of the ceramic material's spec. weight, while that of the blade part amounts to 60-90% of
the theoretical spec. weight of the ceramic material it is made of.

In Patent (220/3) <u>Rolls-Royce (1971) Ltd. (GB)</u> reveal the production
of wheel blades or distributors from a cast piece, one part of which is
made of a powder, composed of a mixture of silicium nitride and a fluxing
agent, which is hot pressed into a matrix, consisting of two sectors,
one of which contains two parts, removable from one another, in order to
selectively compress various zones of the powder during the forming process.

Another Patent (220/2) of <u>Rolls-Royce</u> refers to a composite rotor for
gas turbine motor, provided with an external casing of hot compressed
silicium carbide with at least one heat-resistant element inserted
therein, that essentially fills up the space delimited by the casing,
which is composed of two parts and the cross-sectional thickness of
which is constant in a direction, parallel to the rotation axis of the
rotor. The two parts of the casing contact one another in one point,
while two inserted, heat resistant elements fill up the spaces between
the casing parts, which are joined with one another. The inserted elements are composed of sintered silicium nitride.

Motoren- und Turbinen-Union München GmbH (DE) disclose in Patent (185/1)
a method of encapsulating a body, such as a moulding, of a ceramic material, especially silicon ceramic material, in preparation for hot isostatic pressing, in which a layer of fused silicon applied to the surface
of the material is heated to a temperature of 800-1400°C in a nitrogen
atmosphere to convert the silicon into silicon nitride. The body may be
filled with nitrogen prior to coating so that the conversion reaction
takes place from both the inside and the outside of the silicon layer.
Ductility of the layer is improved by adding 1-10 w.% MgO. The encap-
sulation preserves the original geometrical shape of the ceramic body.
Since a pressure-tight envelope has already been formed around the body
to conform with the contours thereof, before hot isostatic pressing be-
gins and is merely compressed together with it, the risk of deforming
or otherwise damaging the moulding during pressing is eliminated. This
permits the manufacture of complex geometric shapes, such as bladed
wheels for turbomachines. Another advantage is that the encapsulating
layer need not be removed after hot isostatic pressing, because it also
consists of a silicon ceramic material. In a preferred embodiment of
the invention, coating is performed under vacuum to prevent unwanted
chemical reactions during hot isostatic pressing of gas conceivably
trapped in the pores of the ceramic material body.

Motoren- und Turbinen-Union also developed in Patent (185/6) a process
for producing a ceramic turbine wheel by forming a ring by reaction sintering, forming in this ring a disc by hot pressing of ceramic powder
and by fixing, through reaction sintering, to the ring the blades of
the turbine.

The object of Patent (102) of The Garrett Corp. (US) is a ceramic rotor
for a gas turbine, composed of a disc-shaped body of reaction-bonded
silicium nitride with hub part, web-part and bord part, furthermore
with at least one pre-shaped reinforcing ring, made of hot-pressed silicium nitride, which had been embedded in the body, before the reaction
bonding phase and which is monolithically connected therewith. The
reinforcing ring is arranged between the hub part and the bord part of
the rotor.

Patent (98/2) of <u>Ford-Werke A.G. (DE)</u> reveals a process for producing products of double density of silicium nitride, for example parts of gas turbines, wherein the impeller displays a density which differs from that of the hub, connected therewith. A product can be composed of two elements, one of which is shaped from silicium metal particles, through injection-molding,for example, and in which   Si-particles are transformed by adequate treatments into silicium nitride, while another element of the product consists of silicium nitride particles and a compacting adjuvant, these two elements of different density being fitted to one another under pressure.

20.2   I.C. engine parts

Patent (149/1) of <u>Kennecott Corporation (US)</u> relates to a composite sintered ceramic article.which comprises about 5 to about 95 parts by weight of silicon carbide and from about 5 to about 95 parts by weight of titanium diboride. The article, for example in the shape of a honeycomb or a diesel precombustion chamber may be formed by mixing submicron silicon carbide and finely divided titanium diboride in the above proportions with from about 0.5 to about 5.0 parts by weight of carbon or carbon source material,e.g. phenol formaldehyde resin, and from about 0.2 to about 3.0 parts by weight of a sintering aid. This mixture is then formed into a green body,having the shape of the article and sintered under substantially pressureless conditions. The sintered ceramic article displays an electrical resistivity of less than 0.02 ohm.cm and.modulus of rupture greater than 52,000 p.s.i.

Another Patent (149/2) of <u>Kennecott Corporation</u> concerns an improvement of usual internal combustion motors, by producing components of the valve and of the power generating group of sintered SiC, which has been obtained from SiC powder of ultra-fine particles. The structural components comprise valves, valve caps, valve rockers, valve spring holding rings, push bars, cylinders, valve seats, outflow aperture linings, exhaust pipes, pistons, piston rings, connecting rods and the like.

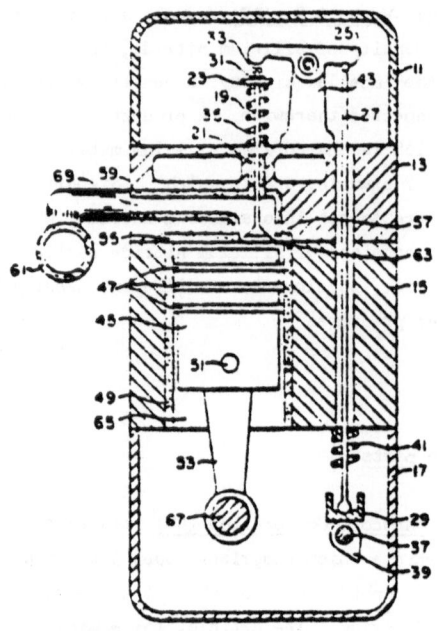

Patent (170/4) of <u>Lucas Industries Ltd. (GB)</u> refers to the production
of a scaling device for a rotary piston internal combustion engine,
comprising a sealing strip having a rubbing surface which includes a
ceramic material,containing at least 90% by volume of a single phase
silicon aluminium oxynitride. The sealing strip includes a silicon
nitride member and the rubbing surface of the sealing strip is defined
by a layer of the ceramic material on the silicon nitride member.

20.3 Machine tool parts

According to Patent (37) of <u>Borax Consolidated Ltd. (GB)</u> granulating
nozzles and cutting instruments are produced from an abrasion resistant
material, which is a mixture of boron carbide (40-60 w.%) and a carbide
or boride of another metal (60-40 w.%), for example of tungsten, titanium,
zirconium or their mixture.

Patent (88) of <u>Edenvale Engineering Works (South Africa)</u> refers to grinding
elements, consisting of bonded abrasive particles of diamond or cubic

boron nitride, the bonding substance being composed of cobalt boride, which is uniformly dispersed in the mass of particles and is applied in an amount of min. 50 w.%. The metallic bonding substance contains 0.5 - 3 w.% boron. The major part of diamond particles consists of synthetic diamond.

In Patent (94/1) Feldmühle A.G. (DE) reveal a cutting plate having improved service life and is made of a sintered material, consisting of 70 to 90 w.% of aluminium oxide, 10 to 30 w.% of zirconium oxide, and 0.1 to 0.5 w.% of magnesium oxide, with not more than 0.6 w.% of any oxidic or other impurities. The material has a porosity of less than 2%, a mean grain size of less than 1.7 μm, and a fracture toughness of at least 190 N/mm$^{3/2}$ at room temperature and at least 140 N/mm$^{3/2}$ at 1000°C.

According to Patent (97/1) of Ford Motor Company (GB) sintered Si_3N_4-based cutting tools are made from a powder mixture of three powders: a first powder of at least 75% crystalline Si_3N_4, a second powder selected from the group Y_2O_3, MgO, CeO_2 and ZrO_2, and a third powder selected from the group of Al_2O_3, WC, WSi_2, W and TiC. The mixture is cold pressed to a density of 50-70% of the theoretical and then sintered to full density without the application of pressure. The third constituent may be included by pick up from the ball milling media used in milling the powders.

Another Patent (97/2) of the Ford Motor Company also refers to the production of cutting tools from silicon nitride by preparing a uniform mixture of silicon nitride powder, having an SiO_2 surface coating, and 4-12 w.% HfO_2 powder; and hot pressing this mixture at a pressure and temperature and for a period of time to achieve a pressed body having substantially full density. Hot pressing is carried out at a pressure of 4-6.5 kst, a temperature of 1680-1710°C and for a time between 1 and 8 hours. The mixture may additionally contain up to 3.5% milling media impurities like W, WC, Al_2O_3 or SiC.

Yet another Patent (97/6) of Ford Motor Company concerns a ceramic material suitable for use in cutting tools and which comprises a hot pressed composition containing Si_3N_4 and at least one of the oxides

$Y_2O_3ZrO_2$ and MgO. Ternary systems based on Si_3N_4, SiO_2 and 4 to 12 wt.% Y_2O_3 are preferred, in which $Y_2O_3:SiO_2$ is from 1.2 to 3.0. Hot pressing at 1700 to 1750°C and 5000 to 6500 psi for 3 to 8.5 hours produces a material with a thermal shock parameter of at least 26×10^9 BUT – lbs hr (in^3) at 1200°C which can be used for making cast iron.

Various types of cutting tool materials have been developed by General Electric Company (US). This Patent (104/3) relates to abrasive crystalline cubic boron nitride, which is nearly completely saturated with phosphor and which can be used in producing a grinding wheel, the external layer thereof being made of particles of the aforementioned boron nitride, which are embedded in a resin matrix, composed of, for example, a phenol polymer. The abrasive particles may be coated with nickel or embedded in a metal matrix .

In Patent (104/6) General Electric Company claim the production of abrasive tools, composed of cubic boron nitride, from hexagonal boron nitride, by mixing hexagonal boron nitride with a powdery metal phase, consisting of iron, nickel, chrome, aluminium, manganese in various combinations. The hexagonal boron nitride content amounts to about 22.w.%.

Patent (104/12) of General Electric Company concerns a drawing ring for fibres, the ring being composed of a mass with a central hole of double-conic shape, the ring mass essentially containing compressed polycrystalline diamond, compressed cubic boron nitride, or their mixtures, the walls of the hole display microroughness and are densely coated with a lubricating agent (graphite, molybdenum, disulfide, hexagonal boron nitride, fats, wax, soap, polytetrafluorethylene) for promoting extrusion.

Patent (104/23) of General Electric Company concerns polycrystalline bodies, which can be produced in a variety of configurations and a wide range of sizes of predetermined shape and dimension. The shaped polycrystalline body is composed of a mass of crystals selected from the group consisting of diamond, cubic boron nitride and combinations thereof, adherently bonded together by a bonding medium, comprised of silicon carbide and elemental silicon, wherein the volume of silicon carbide and silicon each is at least about 1% by volume of said poly-

crystalline body, the crystals ranging in size from submicron up to
about 2000 microns, the volume of the crystals ranging from about 1%
by volume to about but less than 80% by volume of said body, the bonding medium being present in an amount ranging up to about 99% by volume of said body and being distributed at least substantially uniformly
throughout this body, the portion of the bonding medium in contact with
the surface of said crystals being silicon carbide, the body being at
least substantially pore-free. It is useful as an abrasive, a cutting
tool, nozzle or other wear-resistant part.

According to Patent (104/24) of General Electric Company a self-supporting
composite, comprised of a mass of diamond and/or cubic boron nitride
crystals, coated with elemental non-diamond carbon in contact with a substrate, is infiltrated by fluid silicon,producing a likely-shaped composite of a polycrystalline body phase, integrally bonded to a substrate
supporting phase. The composite is useful for the same purposes as that
described in (104/23).

Yet another Patent (104/26) of General Electric Company provides a
process and apparatus for making cubic boron nitride from powdered
hexagonal boron nitride products. Cubic boron nitride is made from
powdered hexagonal boron nitride by a process which comprises vacuum
firing of the HBN and conversion by high pressure-high temperature
processing at 55-80 kilobars and 1600°C to the reconversion temperature.
The high pressure reaction cell has a special design which prevents the
entrance of impurities into the sample. This cell comprises, for example,
a carbon tube enclosing a concentric titanium sleeve. Within the cylinder,
defined by the tube and sleeve are: the HBN sample, carbon filler,
shielding tantalum foil discs and carbon end plugs. The vacuum firing
is done at pressures of 10^{-3}-10^{-10} mm Hg, 1400-1900°C for 5 minutes -
4 hours, and is believed to form a thin, free-boron coating on the HBN
particles. The process works on both pyrolytic (turbostratic) and
graphitic hexagonal boron nitride. Grinding grits formed by milling
cubic boron nitride chunks recovered from the high pressure-high temperature process have resulted in higher grinding ratios than commercially
available CBN.

222

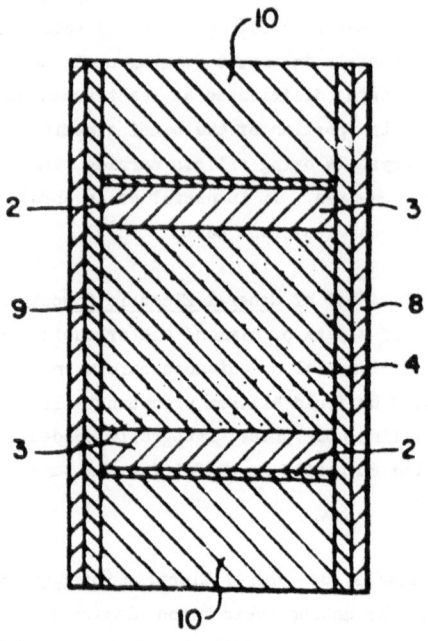

Patent (104/27) of General Electric Company also provides materials for
cutting tools and wear components. According to the patent, sintered
silicon carbide composites, enveloped in diamond or cubic boron nitride
(CBN) crystals are made through a process comprising: (a) forming a first
and second dispersion of uncoated diamond crystals and carbon black in
paraffin; (b) forming a mixture of carbon fibre, carbon black and filler
in paraffin; (c) compacting the dispersions and mixture together to produce an integral compact,wherein the dispersions form an envelope about
the mixture; (d) subjecting the compact to vacuum for a period of time
at a temperature sufficient to vaporise essentially all of the paraffin;
(e) heating silicon to cause liquefaction, direct infiltration and
diffusion into the compact in vacuum, and (f) sintering the compact
containing silicon under conditions sufficient to produce a beta-silicon
carbide binder uniting the composite without applying pressure.
In preferred composites, the dispersions contain different proportions
of diamond or cubic boron nitride crystals. One may be utilised for a
peripheral band about the mixture, the other connecting the edge as a

surface layer of the composition. The composites display extremely high wear resistance.

Patent (116/2) of GTE Laboratories, Inc. refers to abrasion resistant silicon nitride-based articles. According to the patent composite article and cutting tool are prepared by densification to form a body, comprised essentially of particles of hard refractory material, uniformly distributed in a matrix consisting essentially of a first phase and a second phase, the first phase consisting essentially of crystalline silicon nitride and the second phase being an intergranular refractory phase, comprising silicon nitride and a suitable densification aid, selected from the group consisting of yttrium oxide, zirconium oxide, hafnium oxide and the lanthanide rare-earth oxides and mixtures thereof. The hard refractory particles are of an average size below about 20 microns.

Patent (125) of Hoechst A.G. (DE) refers to the production of materials composed of solid, hollow spherical elements, which can be used as catalyst-carriers, destined to the purification of exhaust gases of internal combustion engines, the spherical elements being made of polyethylene, polystyrene, ceramic powder aluminium hydroxide or hydrated aluminium silicates. The spherical elements are coated with an aqueous solution of cellulose ether.

In Patent (170/9) Lucas Industries Ltd. disclose a tip in a rotary turning tool, wherein at least the cutting edge of the tip is formed of hot pressed silicon nitride, a hot pressed ceramic material containing at least 90% by weight of a substituted silicon nitride. The tip in a rotary turning tool includes a body having a cutting edge which extends along at least part of a circle and which is defined between two surfaces of the body, inclined at an acute angle to each other, at least the said cutting edge of the body being formed of hot pressed silicon nitride as herein defined or a hot pressed ceramic material.

Patent (170/14) of Lucas Industries Ltd. provides a lathe tool, attacking a single point, the tool displaying a body with a cutting edge, which is made of sintered silicium nitride or another sintered ceramic material, containing min. 90 w.% substituted silicium nitride. The body may be supplied with several cutting edges, which can be adjusted in

various cutting positions. The sintered ceramic material displays an
average rupture modulus (at $1000^\circ C$) of at least 5980 kg/cm^2. The substituted silicium nitride comprises an oxynitride of silicium and aluminium with the following formula:

$$Si_{6-z} \ Al_z \ O_z \ N_{8-z}$$

wherein z = 0 - 5.

National Research Development Corporation (GB) disclose in Patent (189/2)
a silicium nitride element for application in internal combustion engines,
the element containing a first part in the proximity of a heat source and
is made of a substance of high thermal conductivity, while that part of
the element, which is farther from the heat source, is made of a material
of lower thermal conductivity. A layer of porous or dense silicium of
high thermal conductivity is also arranged over the surface of that part
of the element, which displays high thermal conductivity. The element
displays a thermal conductivity with spatial variations, thus permitting
the reduction of thermal stresses and heat losses. At least one of the
said parts consists of a mixture of silicium powder and a binder, applied
in proportions in accordance with the desired thermal conductivity.

Patent (190/10) of Y.Yoshiro, et al. (JP) refers to ceramic material for
cutting tools, consisting of a sintered, shaped powder mixture, that
contains (a) Al_2O_3; (b) tin and (c) at least one of ZrC, Zr and
ZrN, the component (a) forming 25-80% of the mixture's volume, the volume ratio of component (c) to component (b) being max. 0.5. The
ceramic material is sintered at 1600°-$1700^\circ C$.

Patent (194/4) of Nippon Crucible Co.Ltd. concerns the production of a
tapping tube or stopper bar by mixing together a binder, capable of
forming carbon bonds with a granulated material, consisting of 48-82 w.%
zirconium sand; 10-35 w.% natural graphite (in flakes) and 1-8 w.%
metallic silicium, the mixture being shaped and the shaped product
being sintered in a reducing atmosphere under a pressure between 500
and 1500 kg/cm^2 at 900°-$1200^\circ C$. As a binder tar (of petroleum or coal),
phenol and furane resins are recommended.

Patent (207/2) of N.V.Philips Gloeilampenfabrieken provides a ceramic
insert for a machining tool, consisting of an aluminium-oxide-based
ceramic material, which contains between 10 and 25 w.% of finely-dis-
persed zirconium oxide and not more than 1 w.% of one or more grain growth
inhibitors, the balance being Al_2O_3. The ceramic material has a porosity
of less than 1% by volume. The grain growth inhibiting additive consists
of nickel oxide and/or cobalt oxide in a total quantity of between 0.02
and 0.2 w.% and/or one or more of the oxides of chromium (III), lanthanum
and yttrium in a total quantity of between 0.05 and 1 w.%. The tool has
a long service life.

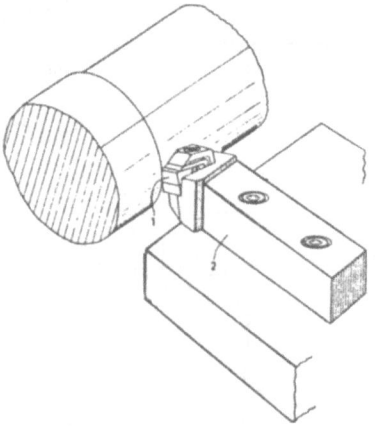

Patent (246/2) of Sumitomo Electric Industries Ltd. (JP) describes
sintered stamping elements for ultrahard tools, the elements consisting
of 20-80 vol.% boron nitride (in high-pressure form), the rest being
composed of Al_2O_3 or a composite ceramic material, which essentially
consists of Al_2O_3, but also contains carbides and nitrides of the metals
of groups IV/a, V/a, VI/a of the P.S., in form of a solid solution or
a mixture thereof, the said rest forming a continuous phase in the
structure of the sintered element.

20.4 Blast furnace and other foundry equipment (components)

According to Patent (5) of Akechi Taikarenga (JP) an immersion nozzle
for continuous casting of molten steel, comprises an alumina-graphite

refractory, having a refractory layer arranged so as to be integral
with the nozzle body and flush with the outer surface of the nozzle
body at that portion of the outside surface of the nozzle body, adapted
to be in contact with a molten mold powder layer on the meniscus of
molten steel, when the lower portion of the immersion nozzle is immersed
into molten steel in a mold, the refractory layer comprising:

carbon (C)	from 2 to 10 w.%
zirconia (ZrO_2)	from 70 to 90 w.%
silicon carbide (SiC) and/or amorphous silica (SiO_2)	from 5 to 27 w.%
and incidental impurities	up to 3 w.%.

(In the figure: 1 = nozzle body; 2 = refractory layer).

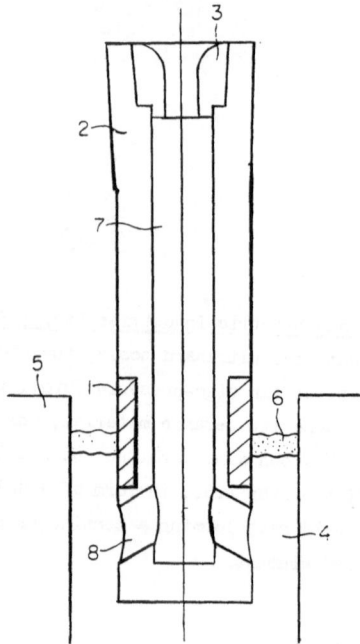

Patent (51/8) of <u>The Carborundum Company</u> refers to an ignition element
of SiC-particles; the element consisting of SiC-particles compacted into
a density of 2.5 g/cm^3. The composition contains negative doping elements

in such an excess of positive doping elements that the spec. volume resistance of the body under cold conditions is less than 1.25 ohm.cm, the ratio between the spec. volume resistance at $20^{\circ}C$ and the spec. volume resistance at $1200^{\circ}C$ being below 12:1.

Nozzles for continuous casting are disclosed in Patent (54/1) of the Centre de Recherches Métallurgiques. A conventional nozzle is made of a composition consisting of graphite (e.g. 30 w.%) and a granulate (e.g. 70 w.%), comprising a major proportion of corundum and alumino-silicates and minor proportions of clay, silicon, and silicon carbide. In order to increase the resistance to chemical attack by a covering powder in the mould, according to the invention at least part of the graphite in the composition is replaced by at least one refractory oxide (e.g. ZrO_2, Al_2O_3) and at least one refractory metal (e.g. Mo, Cr).

Ceraver S.A. provide in Patent (56/1) a material, suitable for the friction parts of, for example, a watch, the material consisting of sintered silicium carbide (beta-form), in an inert atmosphere at a sintering pressure of about 400 bars and at a temperature of about $1950^{\circ}C$, during 10 minutes. Sintering can be effected in a graphite matrix.

Patent (73/2) of Degussa A.G. refers to thermally stabilised aluminium oxide-mixed oxide, obtained by a pyrogenic process and displaying a B.E.T.-surface of 50 to 200 m^2/g. The mixture contains 0.5 to 20 w.% silicium dioxide and due to the presence of this component, it is converted into the alpha-Al_2O_3 phase only at a temperature of $1350^{\circ}C$. The composition is suitable for use in thermal insulation materials.

Patent (74/1) of Denki Kagaku Kogyo K.K. concerns a method for the treatment of ferro-silicon for use as a material in the production of troughs for blast furnaces (iron smelting), ferrosilicon nitride being obtained by blending it with a binder (for example an aqueous solution of polyvinyl alcohol), followed by moulding the blend and nitrogenising the resulting mould, the ferrosilicon thus obtained being soaked in water for at least 24 hours, then removed therefrom and dried. The novel composition is claimed to be superior to conventional clay-bonded trough materials .

Denki Kagaku Kogyo K.K. provide in Patent (74/4) a process for producing ferrosilicium nitride, applicable in the troughs of iron smelting blast furnaces. The said nitride is obtained by mixing 1-100 w.parts of a diluted metal solution and 100 w.parts of ferrosilicium nitride, the mixture being oxidised in a humid air atmosphere, at a temperature between 50 and 200°C. The ferrosilicium nitride can be mixed with mineral acids (HCl, H_2SO_4, HNO_3, H_3PO_4, $H_2Cr_2O_7$, H_2CrO_4), furthermore with organic acids (CH_3COOH, $H_2C_2O_4$, HCOOH, tartaric and citric acid) and also with mineral salts (NaCl, aluminium phosphate, the ferri-salt of chlore and sulphur, aluminium sulphate).

A method, developed by Denki Kagaku Kogyo K.K. in Patent (74/5) for producing ferrosilicium nitride consists of immersing into water ferrosilicium nitride, which had been obtained by smelting together pulverulent ferrosilicium with a bonding agent and nitriding the composition thus obtained. The ferrosilicium nitride, shaped in solid blocks, is suitable for the manufacture of blast furnace troughs.

Patent (147/1) granted to V.P.Kallinin, et al. (USSR) refers to manufacturing refractory products by a method comprising the steps of introducing a mixture of aluminium and magnesium oxides into a mould, with a heater in the central zone of the mould, the mixture being packed into the mould around the heater to form a product; drying the mixture by raising the temperature of the heater to a temperature of 400-450°C at a rate of 30-60°C/min. in an oxidising atmosphere and then to a temperature of 950-1050°C at a rate of 30-100°C/min. in vacuum , firing the product in inert atmosphere by raising the temperature of the heater to a temperature of 1800-1850°C at the rate of at least 100°C/min. and holding it at this temperature for a time sufficient to enable free separation of the above heater from the product and firing it in vacuum at the same temperature. The refractory product is suitable for manufacturing crucibles and pipes of high thermal shock resistance.

A packing mass for blast furnace tap spouts can be made according to Patent (172/1) of Magnesital Feuerfest GmbH (DE) containing 20-60 w.% pyrophyllite (Al_2O_3 . $4SiO_2$. H_2O), 5-30 w.% silicium carbide, 6-20 w.% plastic clay, 5-25 w.% quartzite, 5-15 w.% hard tar (as carbonaceous binder, without elements which become volatile at high temperature) and water for establishing the required density (7%). The proposed mass is adapted to the requirements of the increased capacity of blast furnaces.

Magnesital Feuerfest GmbH also disclose in Patent (172/3) a refractory stamping mass for flues and chutes of blast furnaces, the mass containing: 20-60 w.% pyrophyllite (Al_2O_3 . $4SiO_2$. H_2O), 5-30 w.% silicium carbide, 50-20 w.% plastic clay (Al_2O_3 . SiO_2 . xH_2O), 5-25 w.% quartzite and 5-15 w.% binder, which may be composed of high-molecular tar residue, dry tar or a carbonaceous resin.

Patent (173) of **A. Majdic (DE)** refers to fine-ceramic protective coatings of TiN, ZrB_2, Cr_2O_3, applied by plasma-spraying to ceramic elements, mainly to those consisting of Al_2O_3 (90%), which are used in metallurgical plants and which are exposed to high thermodynamic loads, hot erosion, hot corrosion (discharge chutes, stoppers, closing elements). The coated element is slowly cooled to prevent crack formation in the coating and the substrate. The thin Si-layers applied to the substrate can be nitrided, thereby obtaining Si_3N_4 layers of a given thickness. The nitriding process is intensified by mixing into the Si powder a catalyst, for example iron.

Mannesmann A.G. (DE) reveal in Patent (174/2) refractory material suitable for smelt furnace channels, funnels, containing 20-32 w.% Al_2O_3, 60-70 w.% SiO_2 < 10% of TiO_2; Fe_2O_3, Na_2O, furthermore CaO and MgO in an amount of 1.5-10%. The CaO-MgO component is composed of fine-grained basic blast furnace slag and/or minerals like olivine, magnesite, diopside, hornblende, wollastonite.

According to Patent (184/2), **Moskovskii Khimiko-Tekhnologicheskii Institut Imeni D.I. Mendeleeva (USSR)** developed a method for manufacturing articles from hexagonal boron nitride, which comprises mixing a finely divided powder of hexagonal boron nitride with a boron-nitrogen containing composition, compression-moulding the resulting mixture into blanks at a temperature of from 80 to $160^{\circ}C$ under a pressure sufficient for the penetration of a melt of the composition into the pores of the blank to fill the pores, and heating the blank in ammonia to a temperature of from 1000 to $1400^{\circ}C$, followed by a high-temperature calcination at a higher temperature in an inert atmosphere. The starting mixture can contain 60 to 90 w.% of hexagonal boron nitride and 40 to 10 w.% of the boron-nitrogen containing composition, which may contain boric acid and urea in a ratio of from 1:1 to 1:3. This starting mixture can be compression-moulded under a pressure of from 0.25 to 1.0 t/cm^2 (metric tons/cm^2). The specific surface area for the hexagonal boron nitride powder is preferably from 20 to 300 m^2/g. Using the proposed method, it is possible to manufacture high-purity articles from hexagonal boron nitride, which have increased density and strength, by calcination of pre-moulded blanks. Lower pressures, than have been applied hitherto, can be used during compression-moulding of the blanks. Furthermore, the finished articles have high density and mechanical strength. Articles manufactured according to the invention are refractory and can be in the form of heat-resistant electric insulators, dielectrics in various electronic instruments, as well as an antifriction material, containers for aggressive melts, and the like.

The object of Patent (196/1) of Nippon Kokan K.K. (JP) is a composite sinter with excellent heat and stress resistance, suitable as a nozzle for continuous casting machines, which comprise from 60 to 97 w.% of silicon nitride and from 3 to 40 w.% of boron nitride, present as a dispersed phase in a network of silicon nitride. The composite sinter can be manufactured by (a) kneading from 47.3 to 95.1 w.% of silicon powder and from 4.9 to 52.7 w.% of boron nitride powder as raw materials, using an organic solvent solution, containing a dispersant and a binder; (b) press-forming the resultant kneaded mixture to prepare a green compact; (c) sintering the green compact in a non-oxidising atmosphere at a temperature of from 1100 to 1300°C to prepare a sinter having a strength permitting machining; (d) machining the sinter from step (c) into prescribed dimensions; and (e) sintering again the resultant machined sinter from step (d) in a nitrogen atmosphere at a temperature of from 1250 to 1450°C to nitrify the same. The products made from this material display excellent thermal shock resistance, wear and erosion resistance and allow easy machining.

Patent (196/2) of Nippon Kokan K.K. relates to an immersion nozzle for continuous casting of molten steel, which comprises a nozzle body of an alumina-graphite refractory material and an erosion-resistant refractory layer integral with the nozzle body and flush with the outer surface of

the nozzle body at the outside portion of it, which is in contact when
the lower portion of the immersion nozzle is immersed into molten steel
in a mold, with a molten mold-powder layer on the meniscus of the molten
steel. The erosion-resistant refractory layer comprises from 10.5 to
26.5 w.% of carbon; from 70.0 to 86.0 w.% of zirconium; from 0.5 to
15.0 w.% of silicon and/or ferrosilicon, up to 3 w.% of incidental impurities; and optionally 0.5 to 8.0 w.% of silicon carbide and/or
amorphous silica.

(In the figure: 10 - nozzle body; 11 - refractory layer; 14 - molten
mold powder layer).

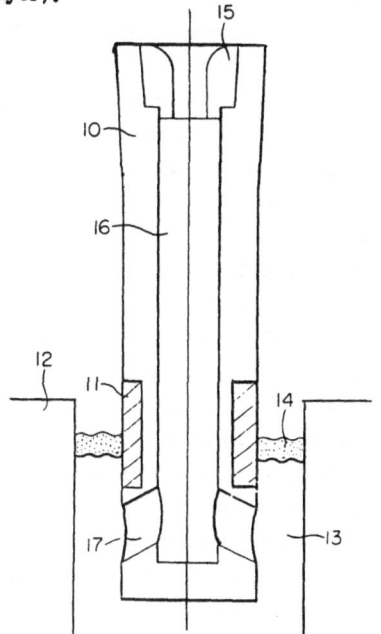

Figure of Patent (196/2)
of Nippon Kokan K.K.

According to Patent (265/4) of <u>Union Carbide Corporation (US)</u> a crucible
is provided, made of pyrolytic boron and suitable for the vaporisation
of aluminium. The crucible is composed of an external wall, an internal
wall and an intermediary wall therebetween. These walls are loosely
joined with one another and the external wall is thicker than the internal and the intermediary walls, while the intermediary wall is not thicker
than the internal one. The crucible is made of boron nitride by a reaction of ammonia and a boron halogenide in vapour phase at 1850°-2100°C

under absolute pressure (not more than 1 mm Hg) for developing boron nitride, which is then deposited on a mandrel, having the shape of the crucible to be formed.

Patent (282) granted to the Vyskumny Ustav Hutnickej Keramiky (Czechoslovakia) concerns a carbon-containing refractory material, including, in a dry condition, from 10 to 99 w.% of calcined anthracite or thermoanthracite or a mixture thereof and from 1 to 50 w.% of a binding agent compatible and capable of being worked with water. Calcined anthracite is anthracite heated at temperatures of more than 1200°C, while thermoanthracite is anthracite heated at temperatures exceeding 700°C but below 1200°C. The material may further comprise, in the dry condition, up to 40 w.% (based on the total weight of the material) of a carbonaceous admixture such as graphite, silicon carbide, foundry or metallurgical coke, or pitch or oil coke, or a mixture thereof, and up to 30 w.% in the dry condition (based on the total weight of the material) of a non-carbonaceous material such as quartz or other silica, aluminosilicate, corundum or a mixture thereof. The non-carbonaceous admixture can in particular be quartz sand, crushed quartz, mullite, corundum or a mixture thereof. The binding agent compatible and capable of being worked with water is suitably clay or solid coal pitch or a mixture thereof or a chemically active binding agent, such as liquid or solid water glass, a liquid or solid phosphate binding agent, such as orthophosphoric acid, a primary aluminium orthophosphate, or a secondary ammonium orthophosphate, either alone or mixed with clay and/or solid coal pitch. The carbon-containing refractory material is intended for making heat-resistant carbonaceous form-pieces and linings for industrial purposes operating at elevated temperatures, particularly in ferrous and non-ferrous metallurgy. The material is particularly suitable for making, repairing and maintaining blast furnace troughs including slag separators, and for repairing and making tapping spouts, siphons and hearths of cold blast and hot blast cupolas. The material can be worked by stamping, vibration and guniting in cold or hot condition.

20.5   Other applications

Patent (1/2) of Advanced Materials Engineering Ltd. concerns a method of making a refractory product of silicon nitride, comprising bonding,

without the use of pressure, a body of green material to itself or to
another body of green material, in which a liquid agent, which wets
the body, is applied to the body or to one or both of the two bodies,
before the body is united with itself or with the other body, the green
material comprising powdered silicon and a binder or binders, to form
a green body and firing the green body in an atmosphere of nitrogen and
at a temperature such that the silicon is converted to silicon nitride.
The said agent may be an aqueous system, containing an emulsion of oil
(silicone or cutting oil) and water, or, as the case may be, a solution
of a surface-active substance. Various products can be made of the
composition, for example regenerative heat exchange discs.

The object of Patent (12/1) of Aluminium Suisse S.A. (CH) is a filter of
ceramic foam of high-temperature resistance, which can be used in filtering
metal smelts and displays an open cell structure. The filter comprises
interconnected cavities, which are surrounded by the foam material. The
density of the filter is 30% lower than the nominal density of the ceramic
material of the same size. The foam contains: 40-95% of Al_2O_3; up to
25% of Cr_2O_3; 0.1-12% of the calcination product of bentonite and/or
kaolin and 2.5-25% of the calcination product of an agent, which is not
reactive with regard to metal smelts and which consists, for example, of
aluminium orthophosphate.

Materials of high heat-absorbing capacity are described in Patent (17)
of B.G.Arabei, et al. (USSR) the material consisting of 12-51 w.% of
boron carbide; 7-22 w.% of silicium carbide; 1-10 w.% of copper;
1-12 w.% of titanium diboride and 79-5 w.% of carbon. The materials display the following properties:
density 2.50 g/cm^3; specific heat over the range of temperatures between 20 and $600°C$, between 0.98 and 1.72 kJ/kg.deg; thermal conductivity over the range of temperatures between 20 and $1000°C$, between 68
and 39 W/m.deg; ultimate bending strength at $20°C$, 12 kg/mm^2; ultimate
compression strength at $20°C$, 24 kg/mm^2 and are suitable for the production of braking devices, for example for air crafts, where the braking
discs may absorb 70 to 98% of the braking energy.

According to Patent (40) of British Iron & Steel Research Association (GB)
the production of a carbon-based article includes the step of mixing to-

gether carbon, a binder, which carbonises to form an amorphous carbon, and a metallic substance comprising nickel, cobalt, or molybdenum, whose ductility is not substantially impaired by reaction with alkali-metals or with carbon, and shaping the said mixture into an article. The binder may be a phenol-hexamine resin. The ductility of the metallic substance is not substantially impaired by reaction with alkali metals or with carbon at a temperature of 300°C to 1100°C and preferably at a temperature of 300°C to 500°C. The carbon-based material may conveniently be produced in the form of a refractory brick.

Patent (49) of <u>Carbomedics ,Inc. (US)</u> relates to the production of all-pyrocarbon prosthetic device components. A complex-shaped component, for example an orifice ring for a heart valve, is made by machining a mirror image of the complex surface in a suitable substrate, such as isotropic graphite. The mirror image of the complex interior surface of an orifice ring is machined in the outer cylindrical surface of a graphite disc, and the machined disc is then coated with pyrocarbon until a coating of the desired thickness has been deposited. After first simply machining away the major portions of the graphite substrate, the remainder of the graphite is removed by abrading with a grit that does not erode the pyrocarbon. After some finish machining on the outer surface of the pyrocarbon deposit and final polishing, the component is ready for use. The production process is illustrated by the following figure:

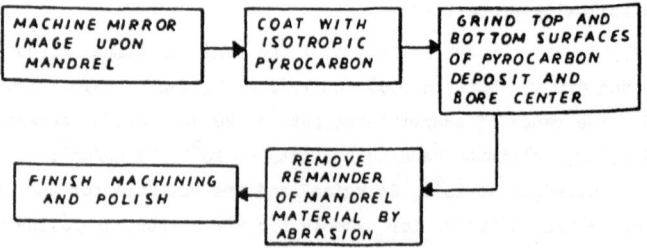

Patent(71/2) of <u>Daimler-Benz A.G.</u> concerns a process for producing
honeycomb-structures, composed of thin separating walls, produced individually from an initial powder mass and then put together to form
a honeycomb green body, which is thereafter exposed to sintering,
thereby bonding together the lamellar separating walls. The structural material consists of a hot-sprayable metal, a metalloid or an
intermetallic composition. This composition is sprayed onto a substrate,
shaped according to the separating wall in the required thickness.
Prior to spraying, the substrate is coated with a separating agent (NaCl).
The materials forming the walls are applied in form of powder in solutions of acryl resin, alkyd resin, in trichloroethylene or methylene-
chloride.

Patent (94/4) of <u>Feldmühle A.G. (DE)</u> provides machine parts of oxide-
ceramic material. A machine part (for example friction bearings and
bearing rings and valve discs) contains 3-25 w.% of zirconium oxide and/or
of hafnium oxide, which are embedded in the other components of the
oxide-ceramic material. The zirconium oxide and hafnium oxide are in
the interior of the machine part in the form of a metastable tetragonal
modification (at room temperature), while at the surface of the machine
part they are present in monocline modification. Preferably the machine
part contains min. 75 w.% aluminium oxide.

<u>Matsushita Electric Industrial Corporation Ltd. (JP)</u> provide in Patent
(175/2) a cantilever, consisting of a tubular body, composed of an inner
layer of crystalline or amorphous boron and at least one outer layer of
amorphous boron. These and other bodies are made by forming at least
one boron layer on a metallic substrate by chemical vapour deposition
and removing the substrate from the boron layer(s) to leave the latter
as a self-supporting body, the substrate removal being effected by an etching
agent comprising a solution (in, for example, an alcohol) of at least
one of bromine, iodine, iodine trichloride, iodine monochloride and
iodine monobromide.

The object of Patent (248) of A.S. Tarabanov, et al. (USSR) is a method
of preparing an antifriction material by forming a mixture of carbon
and a binder, compressing the mixture at a temperature within the range
from 150 to 180°C to produce a blank having a density of 1 to 1.4 g/cm^3,
heating the blank at a temperature within the range from 800 to 1000°C,
and impregnating the heated blank with silicon at a temperature within
the range from 1700 to 2050°C. This antifriction material is stable
under frequent heat variations and retains its properties under oxidation
conditions at elevated temperatures. Preferably the binder is thermosetting resin and the carbon is selected from graphite powder, carbon
black, carbon fibres or mixtures thereof. Impregnation with silicon is
effected in the presence of nickel, cobalt, zirconium, niobium, titanium,
molybdenum, tungsten, tantalum, or chromium, or mixtures thereof and
aluminium and/or iron is added to the mixture during mixing. The material is useful for the production of friction couple elements, such as
scaling rings, thrust bearings, supporting journals, sliding bearings
for electric motors as employed in oil wells; chemical equipment such
as centrifuges and pumps for corrosive liquids, e.g. acids, kerosene,
petroleum, oils, salt solutions, alkali solutions and other liquids which
are corrosive at elevated temperatures.

Patent (249) of TDK Electronics Co. Ltd. (JP) reveals a vessel for
heating food (fish, meat), composed of a sintered body of 4 to 40 w.%
of a ferrite component of the formula MFe_2O_4 in which M is a bivalent
metal selected from Mn, Mg, Zn, Fe, Ni and Cu, and 96 to 60 w.% of a
mineral component of the formula $Li_2O . Al_2O_3. nSiO_2$ in which n is
4 - 8. The sintered body contains clay in an amount of up to 10 w.%.
The method of manufacturing such a vessel comprises: firing at a temperature of from 1050°C to 1300°C for from one to three hours a molding
containing a mixture of 4 to 40 w.% of ferrite powders of the formula
MFe_2O. The described vessel can be made cheaply because natural mineral
such as spodumene or patalite can be used.

P A R T   IV

Equipment used for the production of
ceramic refractory materials and
articles

CHAPTER 21

Structural elements, materials and some special features of equipment

Aluminium Suisse S.A. reveal in Patent (12/2) an apparatus for degasifying and filtering a metal smelt, composed of a chamber with inlet and outlet apertures for the smelt and with walls for supporting at least a first and a second removable filtering element, furthermore a conduit for the feed of a gaseous fluxing agent, arranged between the first and the second filtering element. These filtering elements are made of plates, displaying an air permeability between 400 and 8000 x 10^{-7} cm^2, porosities between 0.80 and 0.95, pore dimensions between 2 and 18 pores per cm and a thickness between 10 and 100 mm.

The Carborundum Company developed a process and equipment mentioned in Patent (51/27) for producing high purity silicon carbide powder (in the

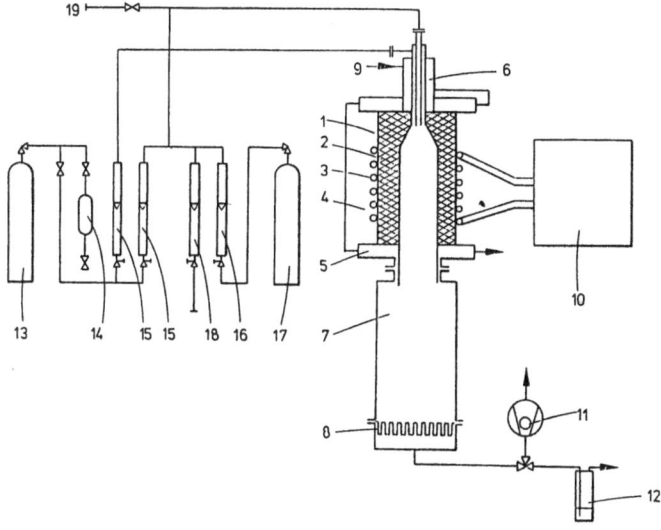

submicronic range) on the basis of alkyl silanes (in the gas phase) by
passing them through a flow reactor, during their passage the gasified
alkyl silane being heated up to a temperature between 450°C and the temperature produced by a plasma burner for realising a thermal scission,
the powder thus obtained being collected in a precipitation chamber.
The apparatus used in this process comprises: a graphite conduit (1)
with conical inlet, surrounded by a thermal insulator; a cooling system
(5) arranged at the end of the graphite conduit; a water-cooled device
allowing to introduce the gas current through a central injector and a
circular injector (6), furthermore a device for the precipitation of the
powder (7).

Patent (63/8) of Chloride Silent Power Ltd. refers to the zonal sintering
of ceramic tubes by passing the tube (or a green body) through a rotary
oven, which is provided (at its input) with elements for entraining the
tube, while imparting to it a mixed longitudinal and rotary movement.
The entraining elements consist of a conveyor and rollers, supporting
the tube to be sintered.

Another Patent of Chloride Silent Power Ltd. (63/13) also describes a
pass-through sintering oven for sintering ceramic tubes. A succession
of tubes, fed lengthwise through the furnace (10, 11, 12) are rotated

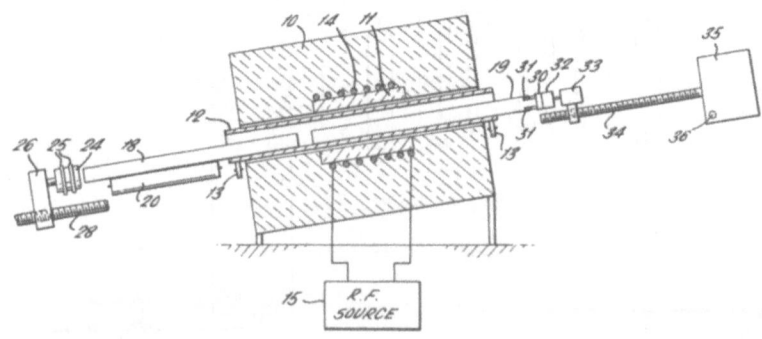

during their passage, the tubes (18, 19), when heated in the furnace
(10, 11, 12), are kept apart by pushing each tube (18, 19) into the furnace at one end with a pusher (29, 25, 26) and pulling it out with a
puller (30, 35) from the other end. The pulling speed is less than the
pushing speed to ensure uniform firing throughout the length of the tube,
allowing for shrinkage as the tube is fired, but is sufficient to hold
the pulled tube clear of the next following tube so as to obviate any
pushing against the end of a hot tube.

According to Patent (66/3) of Compagnie Générale d'Électricité products
of alkali beta-aluminium oxide are produced from a mixture of alkali carbonate powder (sodium carbonate powder) and aluminium oxide by heating
it in an open crucible, followed by cooling and sintering in a tubular
sinter vessel, which is closed at either end with plates, provided with
recesses, which are filled with alkali carbonate, while the lower plate
rests on a bottom and the upper plate is covered with a cover. The
products to be sintered are placed in the interior of the vessel on
blocks, made of the same alkali beta-aluminium oxide as the products.

Another Patent (66/5) of Compagnie Générale d'Électricité refers to a
sinter vessel for sintering alkali beta-aluminium oxide from a mixture
of sodium aluminate powder and alkali carbonate, the mixture being heated
in an open crucible followed by cooling, crushing, shaping and sintering
of the shaped body in a refractory sinter vessel, constructed of a mixture of three components: fire-proof clay of beta-alumina (crushed to
a grain size of about 0.5 mm); a cement (binder) of beta-alumina (obtained from a mixture of alpha-alumina and sodium carbonate) and a sodium salt, in a proportion of 75:20:5. The three-component mixture of
the vessel is made in a moist surrounding by mixing the components at
$80^{\circ}C$ and then firing them at $1200^{\circ}C$.

Deutsche Gold- und Silber-Scheideanstalt and Annawerk Keramische Betriebe
GmbH designed a device (76/4) for producing flat composite bodies, mainly
of silicium nitride, through hot pressing in dies, by exposing at least
one of the preshaped components to a lower temperature than the other one(s).
The space between the heated dies and the component, which has to be exposed to a lower temperature, is filled up with a thermal insulating layer
(of graphite or aluminium oxide felt).

Patent (89/1) of The Electricity Council (GB) relates to a furnace for the
production of tubular or disc-shaped ceramic articles, comprising an open-
ended tube, through which the articles to be heated may be passed, with
heating means for heating the tube, which heating means comprise an induction coil around a susceptor block (of graphite), the open-ended tube
passing through this block; and means for continuously rotating the said
tube about its axis. The furnace has means for forcing air to flow through
the tube, which is at an angle of 4 to 10°. For controlling the temperature
in the furnace, there is provided a second tube through the susceptor block
with open ends, and wherein a radiation pyrometer is arranged for ob-
serving a test element within the second tube. The furnace is also
provided with automatic control means, responsive to the radiation pyrometer and arranged to control power supply to the furnace.

Another Patent (89/2) granted to the Electricity Council reveals an apparatus for producing ceramic beta-aluminium oxide articles, by shaping
them from compressed powder and passing them through a tubular oven, at
the same time passing a current of air threrethrough, which entrains
steam and vapours of sodium oxide, released from the compressed powder.
The air and vapour current moves at the speed of the progressing material or at a higher speed. The gaseous current is realised by forced
circulation.

Heliotronic Forschungs- und Entwicklungsgesellschaft für Solarzellen-
Grundstoffe mbH (DE) designed in Patent (120) an apparatus for the production of silicium rods, the current required for the production process
being supplied by solar cells. According to the patent monocrystalline
silicium is fed from a silicium reservoir into a tubular crystallisation
chamber of graphite, having a vertically removable, cooled bottom plane,
the horizontal dimensions of which correspond to the cross-section of
the rod to be produced. Silicium is continually fed into the apparatus
at such a rate, that the amount of smelt above the crystallising silicium
remains unchanged.

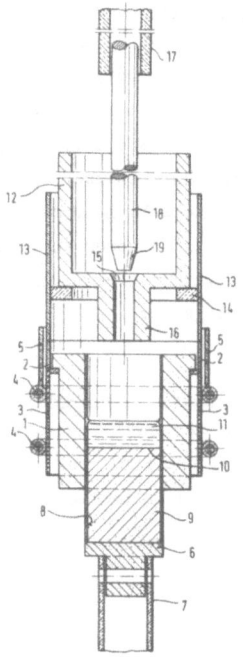

The Bulgarian Institut Po Metalosnanie I Technologia Na Metalite
disclose in Patent (136) a complex installation for nitriding metals
and ferro-alloys under pressure, the installation essentially comprising: a smelting oven, a pulverising device, a tubular oven, a briquetting device and bunkers. The smelting oven is mobile and attached
by a bayonet lock (in its under part) in a feed chamber of the pulverising device, which comprises a crucible, provided with a pulverising
element at its lower face. The feed chamber is fastened to the cover
of the pulverising device, which is connected through apertures in the
cover and an aspirator to a dust separator (cyclone). The lower part
of the pulverising device is in connection with an exchangeable bunker,
which, in turn, is connected to the tubular oven, with a bunker at its
output end, under which is arranged the briquetting device. The smelting
oven, the pulverising device, the tubular oven, the bunkers are in connection with a channel system, feeding in nitrogen and establishing vacuum.

A sintering process for pulverulent, high-density ceramic products, developed by K.K. Kobe Seiko Sho (JP) in Patent (144) can be carried out
in a mobile oven, provided with heating and thermal insulation elements.
The heating oven is introduced in an atmosphere chamber, for presintering
the pulverulent shaped product, while in the atmosphere chamber vacuum
or a predetermined gas atmosphere is established. Thereafter the heating
oven is removed from the atmosphere chamber and then (with its interior still
at high temperature) placed into a high-pressure vessel, wherein the pre-
sintered product (in the oven) is exposed to isostatic hot pressing, at
the same time increasing the temperature of the heating oven, thereby
obtaining a product of high density.

Patent (185/9) of MTU Motoren- und Turbinen-Union refers to a ceramic
firing chamber, provided with air input apertures, the firing chamber
wall consisting, in the zone of the air input apertures, of a double-
walled structure with a hollow space therebetween.

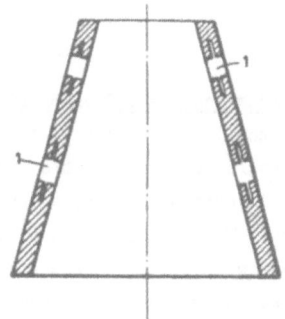

Tokyo Shibaura Denki K.K. disclose in Patent (254/4) an apparatus for
treating powdery materials, utilising microwave plasma, including means
(1; 51) for generating   microwaves, a waveguide (2; 52) connected with
the microwave generating means (1; 51), a reaction vessel (5; 57) disposed
through the waveguide (2; 52), means (11; 120) for supplying the reaction

vessel (5; 57) with a powdery material to be treated, means (19; 123) for supplying the reaction vessel (5; 57) with a reaction gas, and means (14; 136) for exhausting the reaction vessel (5; 57). The reaction vessel comprises a vertically elongated vessel (5; 57) having a bottom above which a plasma generating area is formed, the vessel (5; 57) being so disposed that its central axis is oriented substantially in the direction of gravity. The vertical reaction vessel (57) further comprises a bottom plate (58) horizontally supported in the reaction tube and so provided as to be vertically slidable within the reaction tube. With this apparatus a smaller amount of reaction gas with higher energy efficiency and with yet higher recovery of powdery materials is ensured and a continuous and uniform treatment without any special agitating means is provided.

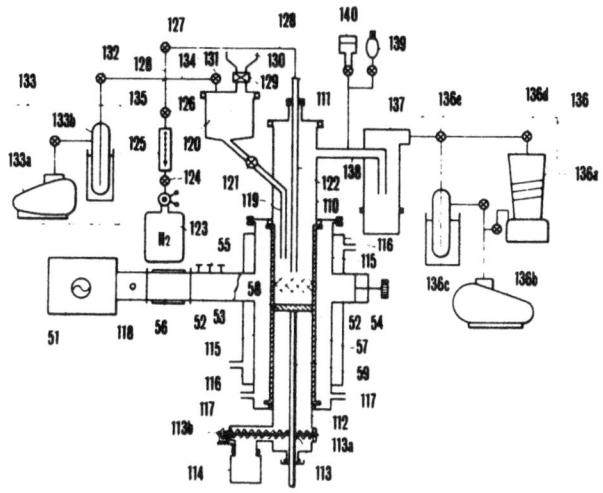

PART V

Miscellaneous

Patent (84/1) of <u>Dynamit Nobel A.G. (DE)</u> concerns a process for the separation of radioactive impurities from crushed baddeleyite, according to which a finely crushed baddeleyite is treated with an aqueous alkali hydroxide solution at a temperature under the boiling point of the solution, during a period of time necessary for the required purification, whereafter zirconium dioxide is isolated. The process may also be carried out by crystallising baddeleyite from a smelt of silicious glass, followed by the separation of zirconium dioxide from the glass smelt.

In Patent (121/4) <u>Deutsche Gold- und Silberscheideanstalt</u> claim a process for producing detergents, cleaning substances, from finely crushed sodium aluminosilicates. The process comprises: the preparation of finely crushed, water-insoluble sodium aluminosilicate, by mixing water-soluble sodium aluminate and water-soluble sodium aluminate in the presence of an excessive amount of caustic soda. The solutions then are fed (at $0 - 103^{o}C$) into a nozzle system with at least two channels, separately from one another, so that the ∨ are united only after having been discharged from the separate nozzles under a pressure of $0.3 - 3$ bars.

Patent (178/1) of <u>Minnesota Mining and Manufacturing Company (US)</u> refers to a method for producing black, sorbent, cross-linked thermoset polymeric foam, which is insoluble in organic solvents and which is obtained by: reacting at its pyrolysis temperature a composition which is liquid at this temperature and comprises at least one aromatic amine (aniline) and sufficient nitric acid to cause initially the evolution of nitrogen oxides followed by a pyrolysis reaction to give an expanded solid which is a black, sorbent, cross-linked, thermoset polymeric foam insoluble in organic solvents.

The purification of waste water is the object of Patent (228/5) of <u>Shell Internationale Research Maatschappij</u>. The patent provides a process for the purification of waste water, in which organic compounds are present, which have a solubility in water at $20^{o}C$ of more than 1 g/100 ml and which contain, in addition to carbon and hydrogen, at least one oxygen atom and/or nitrogen atom, by contacting the water with a crystalline silicate, prepared on the basis of an aqueous mixture, containing the following components:

one or more compounds of an alkali metal and/or an alkaline earth
metal, one or more compounds containing an organic cation (R) or from
which such a cation is formed during the preparation of the silicate,
one or more silicon compounds and one or more iron compounds and in
which mixture the various compounds are present in the following
rations, expressed in moles of the oxides:

$$M_{2/n}O \quad : \quad R_2O \ = \ 0.1 - 2.0,$$
$$R_2O \quad : \quad SiO_2 \ = \ 0.01 - 0.5,$$
$$SiO_2 \quad : \quad Fe_2O_3 > 10,$$
$$H_2O \quad : \quad SiO_2 \ = \ 5 - 50; \quad (n \text{ is the valency of } M)$$

ALPHABETICAL LIST OF APPLICANTS

Page

				Page
14)	American Cyanamid Company	1)	FR-PS 2,351,071 = NL-PA 77,05075	63
		2)	FR-PS 2,397,375 = US-PS 4,119,474	64
15)	Anglian Water Authority		FR-PS 2,376,688	167
16)	Annawerk Keramische Betriebe GmbH	1)	DE-PS 2,361,158	106
		2)	DE-PS 2,412,637	133
		3)	DE-PS 2,458,268	133
		4)	EU-PA 0 045 518	165
		5)	EU-PA 0 052 850	87
		6)	EU-PA 0 052 851	87
17)	Arabei, B.G. et al		GB-PS 1,496,857 = DE-PS 2,637,634	233
18)	Asahi-Dow Limited		EU-PA 0 047 143	19
19)	Asahi Glass Company Ltd.	1)	FR-PS 2,326,388 = DE-PS 2,644,239	5 + 144
		2)	FR-PS 2,394,505	5
		3)	FR-PS 2,478,622	33
		4)	FR-PS 2,486,931	89
		5)	DE-PS 3,006,098	174
		6)	EU-PA 0 063 272	19
		7)	EU-PA 0 070 440	20
20)	Asea AB	1)	FR-PS 2,339,582	107
		2)	FR-PS 2,351,070	107
		3)	FR-PS 2,424,890	107
		4)	FR-PS 2,424,891	108
		5)	FR-PS 2,434,791	108
		6)	FR-PS 2,444,523	175
		7)	DE-PS 3,207,371	108

260

266

268

277

280

281

Page

				Page
272)	VARTA Batterie Aktiengesellschaft	1)	EU-PA 0 046 932	59
		2)	EU-PA 0 067 274	213
		3)	EU-PA 0 068 102	184
273)	VE Wissenschaftlich-technischer Betrieb Keramik		DE-PS 2,711,227	184
274)	Vereinigte Aluminium-Werke AG		EU-PS 0 029 879	43
275)	Vereinigte Grossalmeroder Thonwerke	1)	DE-PS 2,605,949 = FR-PS 2,457,266	132 + 185
		2)	DE-PS 2,605,950	15
276)	Volkswagenwerk AG	1)	DE-PS 2,456,435	25
		2)	DE-PS 2,604,171	214
		3)	FR-PS 2,286,118	75
277)	Vostochny Nauchno-Issledovatelsky I Proektny Institut Ogneupornoi		GB-PS 2,043,048 = DE-PS 2,910,248 = FR-PS 2,455,011	75
278)	Vsesojuzny Nauchno-Issledovatelsky Institut Pozaschite Metallov or Korrozij		GB-PS 2,030,577 = FR-PS 2,440,345	36
279)	Vsesojuznyj Nautschno-Issledovatelsky Institut Abrazivov i Schlifovanija	1)	DE-PS 2,826,544 = FR-PS 2,428,623	39
		2)	FR-PS 2,396,722	160
280)	Vsesojuzny Nauchno-Issledovatelsky i Proektny Institut Tugoplavkikh Metallov i Tverdykh Splavov		FR-PS 2,233,298	39
281)	Vsesojuzny Nauchno Issledovatelsky i Proektnyj Institut Titana		DE-PS 3,005,902	185
282)	Vyskumny Ustav Hutnickej Keramiky		GB-PS 1,485,526	232
283)	Wecht, P.		DE-PS 3,031,526	105
284)	Westinghouse Electric Corporation	1)	EU-PA 0 065 863	185
		2)	EU-PA 0 013 833	213
285)	Wolf, K.		DE-PS 2,358,663	10
286)	Yi Hang Fang		FR-PS 2,390,380	60